Hitherto

A. G. Long mounting cellulose acetate fossil peels in Canada Balsam at the Hancock Museum, Newcastle upon Tyne.
Photograph by Dr Jim Schopf, U.S.A., 1972.

Hitherto

ALBERT G. LONG,
D.Sc., LL.D., F.R.S.E.

The Pentland Press
Edinburgh – Cambridge – Durham – USA

© Albert G. Long, 1996

First published in 1996 by
The Pentland Press Ltd
1 Hutton Close
South Church
Bishop Auckland
Durham

All rights reserved
Unauthorised duplication
contravenes existing laws

ISBN 1-85821-350-9

Typeset by Carnegie Publishing, 18 Maynard St, Preston
Printed and bound in Great Britain by Bookcraft (Bath) Ltd.

This book is dedicated to past and present pupils of the Berwickshire High School, Duns, for their help and inspiration during my teaching from 1945 to 1966.

Contents

	List of illustrations	ix
	Acknowledgements	xiii
	Foreword	xv
Chapter 1.	My Story	1
Chapter 2.	Reminiscences of a Palaeobotanist in Lancashire and Yorkshire	27
Chapter 3.	Berwickshire Nature Notes (1951–1959)	70
Chapter 4.	Entomological Observations	99
Chapter 5.	Articles on General Botany	168
Chapter 6.	Articles on Fossil Plants	185
	Appendix 1	247
	Appendix 2	256
	Tail Piece	262
	Index	264

List of Illustrations

A. G. Long mounting cellulose acetate fossil peels in Canada Balsam at the Hancock Museum, Newcastle upon Tyne. Photograph by Dr Jim Schopf, U.S.A., 1972. . . . . . . *Frontispiece*

Inskip Baptist Chapel, where my father, Albert James Long, was pastor from 1911 to 1919. Photograph by M. Jean Long, 1994. 2

My mother, Isabel Long née Ambler and father Albert James Long. 3

My father at York House, Myerscough, near Preston, Lancashire, the home of Mr and Mrs Thomas Rowe. 4

Family group, July 1924. Jessie Naomi aged 12 (back left); Albert George aged 9 (back right); Marianne Bertha aged 6 (front left); Geoffrey aged 3 (front right). Photograph by Norman Eyre, Todmorden. 6

Headstone of grave of my father and mother at Inskip Baptist Chapel. 11

My brother Geoffrey in R.A.F. uniform shortly before he was killed over Rheims, aged 22, in 1944. 12

Berwickshire High School staff, in the old High School, Newtown Street, Duns, 1946. *Back row, left to right*: Mr Skea (blind Music teacher), A. G. Long, Tom Robertson, Ron Thornton. *Middle row*: Mr W. Brownlie (Janitor), Miss Clark, Miss Rae, Miss Black, Miss Lawson, Miss Mack, Miss Baker, Miss Robson, Miss C. Campbell (Secretary). *Front row*: Mr Ross, Mr Souter, Miss Thorburn, Mr J. Vandore, Dr Imrie (Rector), Miss Mitchell, Mr Clark, Miss Dixon, Mr McNae. . . 14

A. G. Long (left) with Prof. W. S. Lacey (centre) and Dr Arthur Cridland (right) outside my laboratory at the 'University of Gavinton', near Duns, Berwickshire. Photograph taken at the time of the Tenth International Botanical

Congress, Edinburgh, 1964, by Prof. Henry N. Andrews, U.S.A. 15

William Henry Lang, M.B., Ch.B., D.Sc., F.R.S., 1876–1964, my former Professor at Manchester University. Photograph from *Memoirs of the Royal Society*, volume 7 (1961). . 21

Mechanical digger extracting fossil plants from bed of Whiteadder between West Blanerne and Edrom, Berwickshire. . 22

My son David at the White Well, traditionally the source of the Whiteadder above Kingside in the Lammermuir Hills, East Lothian. 23

Manchester University, the entrance I used, on Oxford Road near the Museum where there was a huge *Stigmaria ficoides*, showing to what a gigantic size the arborescent Carboniferous Lycopods grew, as described in E. W. Binney's monograph. Photograph by courtesy of Dennis Print and Publishing from an original by L Price. 28

The seed-fern *Lagenostoma ovoides* Will., three longitudinal sections showing a well-preserved female prothallus with three archegonia around the 'tent-pole', from a coal-ball found on Rowley Tip of Bank Hall Colliery, Burnley, Lancashire, and first described in the *Annals of Botany*. 63

Robin Wood and Whirlaw, from Eagle's Crag, above Todmorden. Photograph by Mr W. Sutcliffe, courtesy of Todmorden Photographic Society. 68

The Retreat, below Abbey St Bathans, Berwickshire, from Cockburn Law. Here I collected moths at night using mercury vapour lamps. Photograph by David G. Long. 84

Gordon Moss, Berwickshire, from the east, another rich locality for collecting moths. Photograph by David G. Long. . . 93

Henry Witham of Lartington (1779–1844). Author of *The Internal Structure of Fossil Vegetables*. Reproduced from *Makers of British Botany*, edited by F. W. Oliver. 95

Gladys and her brother Fred at Hule Moss on Greenlaw Moor, Berwickshire, with the Dirrington Hills in the background. The lake is a roosting place for large numbers of Pink-footed Geese in autumn and spring. 129

View across the River Tweed outside Tweedmouth House,

Illustrations

near the south-west end of the old road-bridge,
Berwick-upon-Tweed.
Photograph by Neil Potts, Berwick 150

My daughter Jean on the cliffs near Lumsdaine, Berwickshire,
with a view south to Pettico Wick and St Abbs Head. 152

W. Ryle Elliot (left), former Secretary of the Berwickshire
Naturalists' Club, with his sister Grace A. Elliot M.B.E.
(centre) and David G. Long (right) on the left bank of the
Tweed below Birgham opposite Carham on the English side. . 177

With Michael E. Braithwaite (left) at the publication of *The
Botanist in Berwickshire* at the Annual General Meeting of the
Berwickshire Naturalists' Club, 1990. Photograph by
Tweeddale Press. 181

Family group (Gladys, David and Jean) at bridge over College
Water, near Kilham, Cheviot Hills, Northumberland. 185

Line diagrams of seeds of the seed-ferns *Genomosperma kidstonii*
Calder and *G. latens* Long. Reproduced from *Transactions of the
Royal Society of Edinburgh*, volume 70 (1977).
Reproduced by courtesy of the Royal Society of Edinburgh. . 186

Langton Burn, Berwickshire, from below the ruined Hanna's
Bridge, 400 yards NNE of the village of Gavinton. Here the
best specimens of fossil seeds of the seed-fern *Genomosperma*
were found in loose blocks washed out of the calciferous
sandstone above the bridge. 187

Dr Peter D. W. Barnard with his father and mother at The
Green, Gavinton, Berwickshire, 1958. 188

The late Professor John Walton of Glasgow University.
From a portrait by Alberto Morrocco, 1962. 192

Trunk of *Pitus withami* from Craigleith Quarry, erected at the
Royal Botanic Garden, Edinburgh. This is Specimen 2 of
Witham found at Craigleith Quarry, Edinburgh, in 1830.
Photograph by Royal Botanic Garden, Edinburgh, and
published in *Transactions of the Royal Society of Edinburgh* (1979).
Reproduced by courtesy of the Royal Society of Edinburgh. . 194

Excavating for compressed fossil plants in the shales at the
Crooked Burn near Foulden Newton, Berwickshire. A. G.
Long (left) with I. McWhan, about 1963/4. Photograph by T.
Huxley, Nature Conservancy, Edinburgh. 196

Thomas Middlemiss Ovens at work on the exposure of shales in the Crooked Burn below Foulden Newton, Berwickshire. Reproduced from *The Border Magazine*, volume 32 (1927). . . . 237

My wife Gladys with Dr H. B. M. Auld, our former much esteemed G.P. at Duns, Berwickshire. Photograph taken on our Golden Wedding Anniversary, 20 June 1992, by Siobhan McDermott. 253

Stoodley Pike, a famous landmark near my home town, Todmorden, first built at the close of the Napoleonic War in 1815 and rebuilt after the Crimean War. Photograph by courtesy of Todmorden Photographic Society. 260

Acknowledgements

It is a pleasure to acknowledge the help received from my patient wife Gladys and from my daughter Margaret Jean and son David Geoffrey. Also from nursing staff at Berwick Infirmary and Tweedmouth House. My thanks are also due to Dr Barry Thomas, Keeper of Botany at the National Museum of Wales, Cardiff and to Miss Helen Fraser for help with typing the manuscript.

My thanks are also due to the members of the Todmorden Photographic Society for permission to use some of their photographs of Todmorden district, also to the Royal Society of Edinburgh for use of photographs from their Transactions. I am likewise grateful to the Editor of *The Scotsman* newspaper who gave me permission to use my nature notes from their column in past copies of their Saturday issues. I wish to acknowledge with thanks much kind help received from Dr C. D. Waterston former Keeper of Geology at the Royal Scottish Museum, Chambers Street, Edinburgh.

Finally, I gratefully acknowledge the generous grant from Scottish Borders' Enterprise towards the publication of this book.

Foreword

This autobiography of Albert Long reveals a life long passion for natural history. From his early life in Yorkshire, through his amateur inerests and professional work, to his retirement years Albert has pursued the subject out of curiosity and for the sake of scientific knowledge. Unfortunately Albert's scientific career started at a very difficult time when war was about to break out in Europe, so life was not to easy for anyone contemplating a career, let alone one in natural history. Following two years in Lang's laboratory in Manchester (1937-39) studying Lancashire fossil plants, he had to go into school teaching. Nevertheless, his interests sustained him and within not too many years he was back studying and publishing on natural history and fossil plants.

Fossil plants have been at the forefront of his scientific work and, quite rightly, honours have come to him. An honorary LL.D. from Glasgow followed fast upon his Manchester D.Sc. and was in turn followed by the Linnean Society Silver medal (for amateur naturalists). But I still remember his modest and somewhat surprised remarks about how recognition and acclaim had suddenly come his way.

Although he is perhaps best known as a fossil plant researcher, Albert has also written widely on botany, insects and other natural history topics. His story is here, wrapped up in a delightful *mélange* of anecdotes, facts about fossils, people and the countryside, and observations upon life. Clearly Albert has been sustained throughout life by a mixture of family, religion and natural history, perhaps the necessary ingredients for a true naturalist.

I first met Albert at the International Botanical Congress at Edinburgh in 1964 when he was already internationally renowned. But it was in 1966 that I first really got to know him when we both moved to Newcastle-upon-Tyne. Albert went to Hancock Museum where he was at last able to spend more than evenings, weekends and holidays working on his important new discoveries, while I took up a Fellowship in the University Botany Department. I valued his company greatly, for he was always ready to talk about fossil plants and freely gave encouragement when it was sorely needed. He could be serious, but dry humour was not too far away even when life was a little difficult. It has been both a pleasure and privilege to have known Albert Long.

Barry A. Thomas.

CHAPTER I

My Story

SINCE coming into retirement, says I to myself, Albert what can you do now? – You have had a stroke, not a stroke of good fortune, but a real stroke, a biological accident caused by a blood clot on the right side of the brain. You cannot walk, but you can talk and you can certainly write right handed. Why not write about your unique experiences and how Providence has guided you? It may help some other pilgrim on life's way. But how do I start, says I? That is obvious, says I, you start at the beginning.

I started life as a fertilised egg as most people do and was born on 28 January 1915 at a little place called Inskip in Lancashire where roses are supposed to be all red. You could pass through this place between Preston and Blackpool without noticing it, but there was a post office kept by a good lady called Mrs Ralph Porter who once gave me stewed blackberries and cream which made me sick in the night so I had to jump out of bed quickly. What Wilfred Pickles would have called an embarrassing moment.

My father was the Baptist pastor at the little Baptist church and I was born in the old Baptist manse (now sold). I lived there until March 1919 when we all moved to Todmorden, Lancashire in the West Riding of Yorkshire. Here my father became minister of the Gospel at Roomfield Baptist church. At the age of 4 years not many memories are registered in the brain but a few still survive. I remember finding a bird's egg on the path round the house, a lilac bush near the gate, a nearby harvest field and hearing the cuckoo, that travelling voice. We had a swing on an apple tree. I recollect sticking a garden fork into my foot, so I must have been prone to accidents. I achieved the feat of running all round the house before breakfast. I have a faint memory of my elder sister Jessie Naomi being ill with scarlet fever, so that we were not allowed into her bedroom but had to look up at her window from the garden. I also remember my mother standing in the road looking for Jessie returning from the village school. Once I was allowed to stay in my father's study while he was preparing a sermon, and my mother enquired how I had behaved and my father said I had been "as good as gold".

Inskip Baptist Chapel, where my father, Albert James Long, was pastor from 1911 to 1919. Photograph by M. Jean Long, 1994.

My birthplace Inskip cast a spell over me and many times through my short life I have returned to it. The rural environment and kindness of the country people together with the fact that my father was buried there in 1940 make it a first place in my life. In 1917 the church celebrated its Centenary by special services in a large tent in the chapel grounds. My father wrote a history of the church entitled "Hitherto" (Ebenezer), and the services were continued annually as the Fylde Convention.

My father, Albert James Long, was born on 28th May 1880 and died on Monday 13th August 1940, aged 60 years. He was born in Cheshire at Torkington Hall near Stockport and died in Leeds Infirmary as a result of a cerebral haemorrhage. My father's parents lived in Gloucestershire, George Long farmer of Walton Cardiff and Marianne Byam Wight of Cheltenham. They were married at Sheepscombe in October 1861. Marianne was only 18 when married and bore nine children of whom father was the youngest. They belonged to the Church of England. George Long was the eldest of three sons, the other two being called John and William. Their father was Samuel Long, who died at the age of 90 in Walton Cardiff. When my father's parents married they set up a dairy farm at Torkington Hall, Cheshire with about £2000 capital. The family consisted of:

1. Frederick William 1862–75 (died Suez)
2. George William 1866 (Canada)

My mother, Isabel Long née Ambler and father Albert James Long.

3. Marianne Elizabeth 1866
4. Helen Annie 1868–1956
5. Edith Mary 1871–1904
6. Dora 1873
7. Arthur Henry 1874–?
8. Katie Louise 1876–7
9. Albert James 1880–1940

My father's eldest brother died tragically of typhoid on a journey home from Jeddah to be married. He was buried in the English churchyard at Suez. Grandfather's farming was brought to a tragic end by foot and mouth disease in his dairy herd. All his cattle had to be slaughtered and buried in quick-lime. This was before the days of Government compensation and it sent him bankrupt. So he moved back to the home farm near Tewkesbury taking one daughter, but grandmother remained in the north of England and took a business eventually finishing up with Auntie Nellie running a guest house at 26 Egerton Road, Blackpool. My father was born after this enforced separation. At first my father went to a private school called Potter's School but subsequently grandmother moved to Lytham and later to Blackpool so his education ceased at the age of 14 years when they had moved to Queen Street in Blackpool. Grandfather died of a growth in his throat, age 52 on the 12th April 1893. Grandmother died in 1920 aged 79 years. Her Daily Light was open on that day at Psalm 116,

My father at York House, Myerscough, near Preston, Lancashire, the home of Mr and Mrs Thomas Rowe.

verse 15. Her favourite passage of Scripture was 1 Corinthians, Chapter 13.

When we moved from Inskip to Todmorden in 1919 I was 4 years of age. My father was concerned about our education. I went to Roomfield Primary School in the centre of Todmorden – a cotton town lying at the junction of three valleys, to Burnley, Rochdale and Halifax. I attended Roomfield Baptist Church in those bleak days following World War I. I remember going to Band of Hope meetings where I signed the pledge to abstain from alcohol when I was 5 years old and by the grace of God I have kept it and never regretted it. On wet Sundays I sometimes stayed to dinner with Mr and Mrs Horsefall who lived in Industrial Street not far from the church.

Things were difficult, so my father resigned the pastorate deciding to go back into business as a bookbinder and to continue preaching the Gospel as a free-lance like Paul who was a tent maker. He had trained in Glasgow at the Bible Training Institute but had not been to an accredited Baptist College and could not, therefore, qualify for the superannuation scheme. He had attended the B.T.I. for five years as a student and assistant (1906–10).

The house where we lived was at 30 Ferney Lee Road and was owned by the Church. Father, therefore, agreed to buy the house with a Building Society mortgage, but had not enough money. To do this he borrowed £150 from Mr Thomas Rowe a cheese merchant in Preston. The loan eased him over the difficulty but the anxiety weighed on his mind. He prayed on his knees for £100. Shortly after (in 1924) he preached at Toxteth Tabernacle in Liverpool and a certain lady called Mrs James came to the morning service and asked one of the deacons if Mr Long was in any special need. But the reply was he did not know. At night this lady handed my father an envelope after the evening service. He put it in his pocket and then forgot about it. On the Monday morning returning to Todmorden on the train he rediscovered it and opened it to find inside a £100 Bank of England note. Later my father returned to Liverpool twice to find the lady and explain his circumstances. She said she had been saving the money for a niece who would probably have wasted it but she was wakened in the night and the Lord told her to help the man coming the next day to preach the Gospel.

When my father repaid the £100 to Mr Rowe he was told, "Cancel the remaining £50". So father wrote in his diary, "All our debts are wiped out." A fine Gospel text. When Mrs James heard the full story she said she was glad to be "the handmaiden of the Lord".

At first my father had a ruling machine in a bedroom and a guillotine in the cellar. Then he got a workshop at Ridge Bank not far from the railway station. Each week-end he was out fulfilling preaching engagements. Below his shop a man used to boil beetroots to sell on the town market. It was a hard struggle for existence through the twenties and thirties. Geoffrey, my brother and my sister Jessie both worked for him. He used to say, "If we go down we will go down with the flag flying." On his shop wall he had the text: "Jehovah Jireh", "The Lord will provide", and He did.

My sister Bertha went to Avery Hill Training College and I studied botany at Manchester University for five years, then went to University College, Hull to take my Cambridge Teacher's Certificate. Victor Murray was my Professor of Education (1939–40). I graduated B.Sc.,

Family group, July 1924. Jessie Naomi aged 12 (back left); Albert George aged 9 (back right); Marianne Bertha aged 6 (front left); Geoffrey aged 3 (front right). Photograph by Norman Eyre, Todmorden.

Class II, div. 1 at Manchester (1937) and then did another year of research for my M.Sc. (1938) on the Fossil Flora of Coal Balls from localities at Todmorden, Bacup, Littleborough, Halifax and Nabb near Water and Burnley. My father's choice of Todmorden proved a good location for preaching the Gospel in Lancashire and Yorkshire and for my finding coal balls, another example of Providence. In 1919–26 I attended Roomfield Council School where the Headmaster of the boys' section was J. W. Crabtree followed by Mr John Briggs son of a previous minister at Roomfield Baptist Church. As boys we played cricket at "playtime" in the school-yard.

My mother was born on 9th July 1886 at Scholes near Holmfirth in Yorkshire. She was Isabel Ambler, married in 1910, and died aged 73 on 29th February 1960. Her mother was Bertha Bateman. Some of her ancestors belonged to the Moravian Brethren.

One day my pals at Ferney Lee Road went along the nearby Lancashire and Yorkshire railway line and came to a little brick building in which they found a small round metal object which they brought to me for identification. It had a flange round the perimeter so I took it into the cellar which had a concrete floor and got the hand axe used for chopping firewood and gave it a sharp blow on the edge to open it and open it did with a mighty explosion. I had made the interesting discovery that the mystery was a fog signal. This brought my mother running down the stairs in a state of fright but relieved to find things all right. So ended my curious experiment with the problem solved and no one any the worse but a little wiser.

In 1940 when I was at Hull I did teaching practice at Goole Grammar School for Boys and I had to go into hospital with appendicitis. It was such a cold winter that in our digs the soap had frozen on the marble washstand in my bedroom so I was glad to get into hospital.

My doctor motored seven miles through the snow to get an R.A.M.C. surgeon who took out my appendix and pickled it for me in a bottle of formalin so I could use it in biology lessons. It had just burst so I had to have a drainage tube. I believe that doctor saved my life by his prompt action, more Providence. He was called Doctor Crawford. The routine in hospital was as follows:

5 am. Wakened, breakfast, egg, bread and butter, tea, temperature taken.

7 am. Day nurses come on; beds made.

About 8.30 am. Sister came round and said "Good morning".

9 am. Dressings, nurse Oldroyd or Sister.

10 am. Matron came round.

I was visited in hospital by my student friend Laurence Gale and by the Biology mistress Miss Tyler. There was a jockey called Glaves in hospital. He had two broken legs. I was kept in bed for nineteen days before I was allowed up. I saw patients being given oxygen and laughing gas, nitrous oxide. One morning was quite sickening because a patient died. Before going he breathed very heavily and finally groaned. I read a book entitled *South* by Shackleton. His boat the *Endurance* sailed into the Weddell Sea and was ice-bound for many months eventually being crushed by ice pressure. An escape was made by three boats to Elephant Island where most of the party remained. Three set out by boat for South Georgia, a distance of 800 miles. Shackleton wrote afterwards,

> When I look back at those days I have no doubt that providence guided us not only across those snow fields but across the storm-white sea that separated Elephant Island from our landing place on South Georgia. I know that during that long and racking march of 36 hours over the un-named mountains and glaciers of South Georgia it seemed to me that we were four not three. I said nothing to my companions on the point but afterwards Worsley said to me "Boss I had a curious feeling on the march that there was another person with us". One feels the dearth of human words the roughness of mortal speech trying to describe things intangible, but a record of our journeys would be incomplete without a reference to a subject very near to our hearts.

All the 24 year age group of students had to register on 9th March, for military service. In the hospital they gave you a dose of cascara if you had not had your bowels moved for two days. I was visited by Professor A. C. Hardy who was Warden at Needler Hall in Cottingham where I stayed as a student. He was a student of marine plankton in relation to fish in the North Sea and invented a continuous recorder which was towed behind trawlers. Later he went to Oxford and took a special interest in religious experiences. Dr Crawford came to see me and asked to see my appendix and said, "What a brute!" On my nineteenth day I got out of bed for the first time. I was up for about ten minutes. It was a peculiar sensation as if the floor was going to come up and hit me. Sister said, "Do not look down, look up." It was strange to look out of the window and see things not seen before though so near.

I had been in hospital before all this because when I was a boy of 14 I was shot in my left foot. I had been with another boy, Jess Barker to the Tennis pond to catch sticklebacks and pond snails for a school aquarium. Coming home to go to the annual Sunday School tea party on 4th January 1930 we saw a man coming down a field. I said to my friend, "Let us see if he has shot anything". I think he must have got over a fence with the gun on cock because when he was down on the path and walking past me only about three or four feet away there was a Big Bang like at the creation of the universe and I felt a stinging pain with a sense of heat in my left foot in the region of the metatarsals. His friend then took the gun and it was as if the Lord put the words into my mouth: I said, "Carry me down that field, there is a man at that big house (Glenroyd) and he has a car". Sure enough the man was in, and the car was in, so forthwith I was taken to our doctor's surgery on Byrom Street. The first thing he did was to give me a jag of tetanus antitoxin. Then I was taken to the ambulance station at Waterside and rushed to Halifax Infirmary (twelve long miles) with a tourniquet on my thigh. How glad I was to get under the anaesthetic – the old A.E.C. mixture drop by drop until I was out for the count. What a relief. When I came round I had a cage over my foot and a drip of hydrogen peroxide to keep the wound free from infection. For two months I was in Halifax Infirmary, bed No. 22 in Porter Ward. I thought I would never climb over a dry stone dyke again, but I did. I raced along the corridor in a wheelchair and pulled up so quickly that I shot out of it and landed on the floor none the worse. But sister was not exactly pleased.

After two months I went home and was on crutches for eighteen months. Then I got a surgical left boot and learned to walk again. One result is that my left leg is one and a half inches shorter than my right leg so that is the long and short of me. When my father first heard of my accident, he said, "There will be a providence in it", being a man of faith; and so there was, as events proved, because when I was medically examined for military service I was medical Grade 4 and so escaped the awful business of shooting others in World War II. This meant I could take a teaching post and help my mother financially at the outbreak of the War. One legacy of my teaching practice in Goole was that Gale introduced me to Shakespeare's sonnets, for example:

> Let me not to the marriage of true minds
> Admit impediments. Love is not love
> which alters when it alteration finds,
> Or bends with the remover to remove:
> O no; it is an ever-fixed mark,
> That looks on tempests, and is never shaken;
> It is the star to every wandering bark,
> Whose worth's unknown, although his height be taken.
> Love's not Time's fool, though rosy lips and cheeks
> Within his bending sickle's compass come;
> Love alters not with his brief hours and weeks,
> But bears it out even to the edge of doom.

If this be error, and upon me prov'd,
I never writ, nor no man ever lov'd.

While in hospital I read a life of Cromwell by Harrison. Cromwell lost his son Robert at the age of 18 years. Nineteen years afterwards, fresh from his daughter's deathbed, almost on his own deathbed, the broken hearted father recurred again to this his first great loss. He had read to him those verses in Philippians which end "I can do all things through Christ which strengthens me" (Phil. 4: 13), and then he said "This scripture did once save my life, when my eldest son died, which went to my heart like a dagger, indeed it did". Death is a strange and potent reality. How significant is it that the Lord had power over both disease and death.

When I went back to Needler Hall at Cottingham the taxi man forgot to turn up, and I nearly missed my train. Prof. Hardy met me at Paragon Station. Back at Needler Hall the matron at first insisted I should have breakfast in bed; but this I never liked so was soon on my two feet again.

On 29th February 1940 the college doctor said I should be ready for home in three days. By March I was at Cottingham again and returned to Goole on 5th March 1940 to complete my teaching practice. At night before dark Laurence and I went down to the docks and we watched a number of tugs being moored. I had a letter from my mother and father and my mother was very displeased with me for going back so soon. However, it was worth it as I caught a melanic Pale Brindled Beauty moth on a horsechestnut tree in Needler Hall grounds.

After completing my teaching practice in early 1940 I had to seek a job. I applied for a temporary post at Lewes in East Sussex and went down for interview in June. It was just at the time when our troops had escaped from Hitler's maw at Dunkirk. There had been a special national day of prayer and God answered it. I was offered the post and in faith I took it though I felt I might be entering Hitler's mouth. He was just at the other side of the Channel. My father died in August so the post came in the nick of time as it meant I could help my widowed mother to repay the mortgage on the house. I found Lewes to be a lovely place to live and was greatly taken with the South Downs where I first saw Noctule bats flying in the sunshine at evening. I also remember seeing a Red Underwing moth sitting on a telegraph pole. We had an open air swimming pool at the County School for Boys and I saw a grass snake swim a breadth which was about all I learned to do. It was the time of the Battle of Britain and German bombers would turn back to the Continent and jettison their bombs anywhere. I stayed at 33 St. Swithun's Terrace near a big house called The Grange and one Saturday morning a German plane dropped a stick of bombs over the town and I heard one coming down with a swish. I expected a mighty explosion but instead it buried itself in the ground. I hid myself near the fireplace and when I looked in the mirror my face was as white as death. Providence again.

Before the death of my father he took me over to Bacup to preach at Doals Ebenezer Chapel at Weir village between Bacup and Burnley, 1,000 feet above sea level. He asked me to read the lesson which I did. At lunch time we went to the home of the Secretary Mr Henry Hunt. There I first met my future wife Gladys. It was love at first sight but it cost her half of her chop for dinner as I was invited to stay. My father knew nothing about this romance, and being a cautious lad I prayed a lot about it. I asked her if she could wait five years before we got married but I could only wait eighteen months. My father had died on 13th August 1940 and was laid to rest at Inskip Baptist Chapel-yard 16th August. Bertha chose for a memorial verse.

> Love ran through his life like a thread of gold.
> And the life bore fruit a hundred-fold.

Gladys and I were married at Doals Church in June 1942. By this time I had secured a permanent teaching post at Leek High School for boys in North Staffordshire near the southern end of the Pennine

Headstone of grave of my father and mother at Inskip Baptist Chapel.

chain. Here I stayed for three years and was becoming quite fond of the place, especially of the hills between Leek and Buxton. There was a wild place called Threeshire Heads where Staffordshire, Derbyshire and Cheshire all met. I preached the Gospel in over twenty little Methodist chapels in the Leek and Moorlands Circuit as well as at open-air camp meetings.

On the night of 3–4 May my brother Geoffrey was killed over Rheims in France. He was a wireless operator on a Lancaster bomber. It was a moonlight night and the Germans switched their fighter planes to the west. Only one of the eight crew members was saved. He was the rear gunner and had an ejector seat. The last thing he saw was the plane going down in flames. Geoffrey was only 22 years old and was the only one of us born in Todmorden.

My brother Geoffrey in R.A.F. uniform shortly before he was killed over Rheims, aged 22, in 1944.

I was not settled in the school at Leek so began to look out for another post. I applied for a post at Berwick High School but failed to get an interview then a vacancy occurred at the Berwickshire High School in Duns so I applied for it and was invited to go for interview in November 1944. We did not know where the place was and had to look it up on a map.

On the journey north everything went wrong. There had been a smash on the railway near Crewe, so I was delayed and missed my connection at Carlisle. I slept that night in a station waiting room with a lot of soldiers and airmen and a Glasgow diver whose language was most lurid. On the Saturday morning I got a train for Duns but had to change at St. Boswells. I was travelling with a soldier to Greenlaw and he directed me to the train for Duns but it never got there. Instead I found myself at Kelso and had to telephone the Director of Education at Duns. So I had to get a bus to Duns. I first had some breakfast and then caught a bus. After all this mixup I arrived at

Southfield House and was offered the post. I had had such a lot of trouble getting there that I felt I would refuse it but I accepted it and so began my sojourn in Scotland. It proved to be another example of Providence.

There were a lot of Polish soldiers in Duns at that time and it was difficult to get digs as the Poles got the first chance. I stood at the foot of Bridgend and cogitated. I wondered whether to abandon the whole business for I had given up my half term holiday to come to Duns and seek digs. Dr Cairns was Chairman of the Education committee and he took me to No. 5 Bridgend and there the good lady took pity on me so at last we got a roof over our heads. I have always remembered what Dr Cairns did. He gave me three eggs in the time when we had rationing and said, "We cannot let you go back to England without some of the grapes of Eshcol." So when I got back to Todmorden I gave one to Jessie and she ate it for her breakfast, fried. What a treat! By the time I got back to Leek my wife said I looked like a tramp because I had never shaved on my travels.

But it all worked out and before very long we were moving again. Seven times in seven years if I remember aright. First to Gourlay's Wynd and then to Preston Schoolhouse. This was our first home of our own, between Duns and Grantshouse on the A1. We were very happy there and in 1947 Gladys had to go into the Knoll at Duns, the most expensive hospital in Scotland, where our daughter Margaret Jean was born. As soon as that occurred it began to snow on 19th January and it continued into April so that at one time it was up to the top of the telegraph poles. Her confinement was on a Sunday so the good lady who attended her said she did not like Sunday confinements, she liked to go to church. My wife said she was not a teacher and did not want her own way. Next time she would try better, but fifteen months later David Geoffrey came and would you believe it? It was on another Sunday, poetic justice; Providence again.

Before long we were on the move again. It was to Gavinton between Duns and Greenlaw where we had two cottages made into one at the north-east corner of the Green. This was a place of much domestic bliss although there was not a straight wall in the place. I knocked down a wooden shed in the garden and had the wash-house rebuilt as a honey house where I could extract honey. I had started keeping bees at Preston with some hives in Bonkyl House garden. At Gavinton I had an apiary in Langton estate and at Kyles Hill. This was a marginal site yielding both clover and heather honey. On the title deeds of our cottage we had the signature of Gavin, the factor at Langton who gave his name to the village. Our move to Gavinton was another act

Berwickshire High School staff, in the old High School, Newtown Street, Duns, 1946. *Back row, left to right*: Mr Skea (blind Music teacher), A. G. Long, Tom Robertson, Ron Thornton. *Middle row*: Mr W. Brownlie (Janitor), Miss Clark, Miss Rae, Miss Black, Miss Lawson, Miss Mack, Miss Baker, Miss Robson, Miss C. Campbell (Secretary). *Front row*: Mr Ross, Mr Souter, Miss Thorburn, Mr J. Vandore, Dr Imrie (Rector), Miss Mitchell, Mr Clark, Miss Dixon, Mr McNae.

of Providence because Gladys's father died and we were able to offer a home to her mother and care for her. She lived with us until she died on her ninety-sixth birthday at Ponteland. Another aspect of Providence was that we were very near the Langton Burn. Here there was a broken down bridge known as Hanna's Bridge where I later found some of my most interesting fossil plants.

I had become interested in fossil plants when at Manchester University through another student from Heywood and we had collected together on old tips of the Upper Foot seam at Dulesgate between Todmorden and Bacup. I used to combine courting with collecting coal-balls in which I found a good specimen of the seed *Lagenostoma ovoides* with a perfect female prothallus showing three archegonia and a tent-pole almost like *Ginkgo*.

My honey house at Gavinton became my laboratory which now

A. G. Long (left) with Prof. W. S. Lacey (centre) and Dr Arthur Cridland (right) outside my laboratory at the 'University of Gavinton', near Duns, Berwickshire. Photograph taken at the time of the Tenth International Botanical Congress, Edinburgh, 1964, by Prof. Henry N. Andrews, U.S.A.

served as the "University of Gavinton". Here I made many sections of the blocks first found in the Langton Burn by the new cellulose acetate peel technique which came just at the right time for me to use it. Another Providence for me was the big flood of 1948 which scoured the bed of the Whiteadder and brought to light many good things. I used to walk about with fishing waders on and a miner's pick in my right hand to turn over the boulders to see if there were any fossils there. In this way I made some very interesting discoveries especially of pteridosperm seeds. These were described in the *Transactions of the Royal Society of Edinburgh*. My two first papers were in the Annals of Botany and drew forth a letter from F. W. Oliver in Egypt. He and D. H. Scott had first proved that many of the so-called ferns in the coal measures were not ferns at all as they bore seeds. So they were called Pteridosperms or seed ferns and now many of us think they were the ancestors of the modern flowering plants known as Angiosperms.

I got much help and encouragement from Professor John Walton of Glasgow University. But the man who helped me most was a young student of Birkbeck College, London, who was paralysed from the waist downward by an accident in the gymnasium at school. He was called Peter D. W. Barnard and wrote to me out of the blue in early 1957. It was he who jolted me back into palaeobotany. I had become a bit disillusioned and turned back to my first love of entomology, keeping bees and collecting butterflies and moths then caddis flies. The results were published in the *History of the Berwickshire Naturalists' Club*, the oldest field club in Britain founded in 1831 like the British Association for the Advancement of Science. In 1966 I put my research papers on palaeobotany forward, bound in triplicate, for the degree of D.Sc. at Manchester University and was successful. Glasgow University gave me the honorary degree of LL.D. so in 1967 we had to attend two graduation sessions something I had never dreamed of. Secretly I would have preferred to remain plain mister. In 1966 I had become very disillusioned as a biology teacher at the Berwickshire High School so I reluctantly decided to look out for another post. I wrote to John Burnett, the Professor of Botany at Newcastle University, and asked if there were any vacancies in his department. He sent me a very gracious reply and informed me in the negative but said there was a vacancy in the Hancock Museum. Eventually I went for interview and was offered the post which I accepted. When I left the Berwickshire High School in 1966 the pupils made me a presentation after morning assembly and I made a parting speech as follows:

In making me the recipient of this kind gift you have honoured me more than I deserve and at the same time have set me the problem of how best to thank you.

When I came to Duns on New Year's Day 1945 I was a relatively young man in the prime of life. I have therefore served my apprenticeship 3 times over in the B.H.S. Twenty-one years is a rather tidy figure 3 × 7 – the perfect number and a good slice out of anyone's life.

One of the first things that struck me when I came to Duns was the clean and natural colours of the trees and the good growth of lichens. This was a sure sign of purity, purity of the atmosphere.

I lodged at the foot of Bridgend not far from the old cattle market and that winter was so cold that my breath froze on the blanket in bed. We had a very cheery janitor at the old High School called Watty Brownlie. He was the first person to welcome me at the school with a face beaming like the rising sun.

Where we lodged we had no bathroom so I went to the Waverley Hotel for baths. It was kept by a kind lady called Mrs Jackson and when I asked her what the charge was she said there was no charge but she had a missionary box in the dining room so I used to drop something in that.

In those days it was possible to travel to and from Duns by the railway. Once when I was going to the station carrying a large suitcase a post van drew up and the postman said, "Could I take it down to the station for you?"

Kindnesses do not always cost much but they make a great difference in life. I therefore treasure your kindness to me just as the gift you have given me. Education is no good if it makes people greedy or selfish or unjust. It should cure these evils. All the certificates in the world therefore cannot be compared to a kind spirit.

You who are leaving B.H.S., as I am, will find that there is much greed and selfishness in the world, but there is also something else. Here and there you will meet with true kindness. Jesus said that God was even kind to the unthankful. The Bible calls that the grace of God. This kindness of God is therefore like a light from heaven. At times the world is very dark and the way is very rough but Jesus said, "I am the light of the world. He that followeth me shall not walk in darkness but shall have the light of life". Have you become a follower yet? The first step is to let the kindness of God come into your life. It will give you new hope and new faith. Your kindness to me gives me new faith. We live by faith. We all need faith. It is the spring of action. I shall therefore complete my thanks to you by quoting an old song which I think is suitable for me the receiver and for you the givers:

> Have you had a kindness shown? Pass it on.
> 'Twas not given for thee alone. Pass it on.
> Let it travel down the years.
> Let it wipe another's tears,
> Till in heaven the deed appears. Pass it on.
>
> Be not selfish in thy greed,
> Pass it on.
> Look upon thy brother's need,
> Pass it on.
> Live for self, you live in vain,
> Live for Christ, you live again,
> Live for Him, with Him you reign.
> Pass it on.

And don't forget if you are passing through Newcastle just drop into the Hancock Museum and search me out among the other curios.

I really did not want to leave Gavinton. It was like going from heaven to hell as I am not a city man. However, all worked out for my good, Providence again. I stepped down in my salary but eventually came out better off. Moreover I could continue research in the back corridor of the museum and was helped greatly by the Curator who kindly saw that I was provided with rock-cutting machines and other necessary apparatus and reagents. Furthermore, I found I could motor northwards to the lovely Scottish Borders in my Morris Traveller. I also motored down to Todmorden and Burnley and brought back coal-balls to Newcastle, many hundreds of them. Sometimes I wondered if the Hancock Museum would sink at my end of the back corridor but the foundations stood firm being Victorian. Forty years before while at Manchester I had prayed for a *Lagenostoma* seed cut longitudinally, and now after all those years God answered my prayer with a specimen showing three archegonia round the "tent pole". But of all my finds I still was most impressed with *Genomosperma* and *Stamnostoma* from Berwickshire. They supported the theory that the seed integument was derived from telomes. Before long I was on the scent of another theory namely that the carpel in Angiosperms was possibly derived from the cupule in Pteridosperms. But this is another exciting story. Much more evidence is needed.

When we moved to Newcastle we bought a terraced house in Jesmond at No. 26 Cavendish Place within a mile of the museum so that I could walk to work and keep fit. It was here I first heard collared doves chanting on chimney pots, "Nasser is dead", but unfortunately for us Nasser was not dead as we found out to our cost. After some time I began to think I had made a mistake in moving to Newcastle and I became very restless. However, I met a lady on the street and voiced my feelings to her. She was a former Methodist minister's daughter and said to me, "Give yourself time or you will be running round in circles." This was just the word I needed and felt it was Providence again. I had thought of applying for a teaching post in Hawick; it would have been sheer madness.

However, the bricks and mortar of Jesmond got too much for me, so we decided to flit again, and bought a house at Ponteland: 64 Cheviot View. We arrived just when the laburnums were in bloom. It was nice to see green fields and soon I was motoring the sixty miles journey northwards to collect fossil plants in Berwickshire and up the East Lothian coast. I used to go to Oxroad Bay near Tantallon Castle and search the shingle with Wellington boots on. I found the same species as in the Cementstones of Berwickshire together with some that did not occur there. I got much help from Dr C. D. Waterston of the Royal Scottish Museum. One of the most interesting discoveries was the evidence that the huge fossil tree known as *Pitus* was probably a Pteridosperm. In this connection I was able to motor from Ponteland to near Gilsland along the military road near the Roman Wall of Hadrian. Here, near where the Trout Beck joins the King Water, a tributary of the River Irthing, there are some large fossil tree stumps of *Pitus* in the river bed. It is near Spadeadam Waste where the old Blue Streak rocket range was. I nearly got arrested by a Ministry of Defence officer for taking photographs of the fossil tree stumps. He was on the defensive for fear of the I.R.A. So I invited him to come and have a fly cup of tea at the Hancock Museum in Newcastle. I once took Professor Henry Andrews to see these fossil *Pitus* tree-stumps on a very cold day near the end of January. He never forgot it.

Eventually, when I became 65 years of age, I had to retire from my post at the Hancock Museum. In order to continue my research on fossil plants I came to Berwick-upon-Tweed so my life has gone full circle from wheelchair to wheelchair but God has never deserted me and I have made many new friends, and have had much kindness shown to me especially by the nurses in the Berwick Infirmary. Now

I have flitted again to a house near the west end of the old road bridge where I am very comfortable and well cared for. The problem now is to redeem the time as profitably as is possible.

My professor at Manchester University was W. H. Lang who along with Robert Kidston of Stirling published some classic papers on silicified fossil plants at Rhynie in Aberdeenshire. I visited the site in 1964 with the Tenth International Botanical Conference held in Edinburgh. The chert bed was opened up by A. G. Lyon with help from a special grant from the Royal Society of London. It was a great success and greatly impressed our American friends. Soon after I came north to Berwickshire I visited Cockburnspath on my bicycle and discovered some compressions in a bed of shale near Cove Harbour. I sent some to Professor Lang at the Manchester Museum and asked him what geological age they were. I did not have a geological map but he looked up the old one by Geikie and told me they were Lower Carboniferous and I would have some fun studying them. How right he was.

Later I revisited this place and it gave me some very good petrifications like those in the Langton Burn near Gavinton and the Whiteadder near Hutton Mill, Chirnside, and West Blanerne. Here I was helped by the farmer Mr W. Pate who allowed me to get a bulldozer to excavate the river bed. I used to use a crow-bar to dislodge the large concretions in a bed of shale then break them with a 14-lb sledge hammer and carry them up to the farm or to Edrom Mains. Then they were taken down to Newcastle to be cut on the slab saw into slices 1cm thick. We used paraffin and flushing oil as a lubricant so the slices had to be washed with detergent and left in a hot air oven overnight. Then peel sections were made and the thickness of the slices measured by vernier calipers.

In this way reconstructions of the fossil plants were made. Such reconstructions included the internal anatomy and were drawn in the round. This had been done for the coal-ball fossils especially for *Botryopteris hirsuta* which showed small plants with roots borne on foliar organs; a method of vegetative reproduction as in modern ferns such as *Asplenium bulbiferum*. It is possible that *B. hirsuta* was an epiphyte on the giant trees of *Lepidodendron*. Some of these arborescent lycopods also bore seed-like organs known as *Lepidocarpon*. It was interesting that Kidston and Professor Gwynne-Vaughan had commenced work on the fossil flora of Berwickshire in a paper on *Stenomyelon* which came to light when a drain was dug across the road north of Norham Bridge. After the death of Professor Gwynne-Vaughan, Professor Lang went to Stirling to see if the work could be continued. However, by

William Henry Lang, M.B., Ch.B., D.Sc., F.R.S., 1876–1964, my former
Professor at Manchester University.
Photograph from *Memoirs of the Royal Society*, volume 7 (1961).

Mechanical digger extracting fossil plants from bed of Whiteadder between West Blanerne and Edrom, Berwickshire.

that time the Rhynie fossils had been discovered by Dr Mackie who was Medical Officer for Health at Elgin and a keen amateur geologist. He was sitting on a dry-stone dyke at Rhynie eating his sandwiches when he saw a peculiar piece of chert with spots in it. Dr Kidston was honorary palaeobotanist to the Geological Survey in Scotland and pursued this discovery so the fossil flora of Berwickshire was left until, in the providence of God, I came along and at Gavinton found a field of research just on my doorstep.

The great flood of 1948 was also a providence to me as it left behind huge shingle beds. Moreover, the bed of the Whiteadder and Blackadder were bulldozed afterwards and just above the Blackadder bridge at Allanton a nice pile of flat concretions had been left just waiting for me to collect them. It was near here in an old quarry that Henry Witham found his *Anabathra pulcherrima*. On a shingle bed above Hutton Mill I found a block from which I made over 900 peel sections which gave me the new seed of *Stamnostoma* and its beautiful petrified cupule. It is fortunate there was no loss of life in that terrible flood. We were at Preston at that time and I helped to clean out two of the cottages at Cumledge Mill.

I first set eyes on the Whiteadder at a place called Marden and was so impressed by it that I said to myself I would like to trace this river to its source. Later I managed to do this and took photographs of it from source to its mouth where it runs into the Tweed opposite East

My son David at the White Well, traditionally the source of the Whiteadder above Kingside in the Lammermuir Hills, East Lothian.

Ord. In this way I got to know it very well and this was useful when I came to look for fossils and catch lepidoptera.

I learned to drive when I was 39 years old. I bought an old post office van for £68 and had windows put in then it was painted green. I got an old motor car seat in the back. We had some exciting moments with it. Having only a 6-volt battery it sometimes refused to start. Twice the foot brake refused to work. Once it was up near the Retreat when I was catching moths at night. I was never very good at reversing and when turning bumped into a tree. Below was a great drop down to the Whiteadder so I was scared but managed to hold onto the hand brake. The other time I had taken two pupils from Ayton up onto Coldingham Moor for a night of mothing. I had to get them home in the morning by driving all the way in second gear. Once the gasket blew when I was up the Hungry Snout and the engine chugged like a steam roller when I was going up the Stony Moor approaching Duns. I left the old banger outside the Swan Garage and walked the three miles back to Gavinton.

Once when I was working my mercury vapour light trap at the Hirsel the lady at the big house saw my lights and must have thought I was poaching so she rang the Coldstream police and a policeman came on a motorbike. I thought it was the fire brigade, but he was much relieved when he found I was only catching moths. Unfortunately, he went and knocked up the gamekeeper at about 2.30 a.m.

and he had to jump out of bed and come downstairs in his pyjamas so you can guess how popular I was. But it was well worth it as I caught a Convolvulus Hawk Moth, a female, but she had laid all her eggs. Back at Duns I went round the street lamps and found another, a perfect male twenty feet up a lamp standard outside the Horn Inn. It is now in the Hancock Museum.

I managed to pass my driving test and then bought a second-hand Morris Traveller which we used when we flitted from Gavinton to Newcastle. What a flitting that was! A day much to be remembered, like going from heaven to hell. But it all worked out, as the Lord said to Paul when he prayed about his thorn in the flesh: "My grace is sufficient for thee, for my strength is made perfect in weakness".

I treacled all the telegraph poles between Duns and Longformacus using my bicycle as transport. Other places I visited by night were Gordon Moss, Kyles Hill, Siccar Quarry, Marden and Duns Castle Lake and Oxendean Pond where an otter came up the burn whistling then dived into the pond and scattered the wild ducks roosting on it. At Kyles Hill and by the A1 near Grantshouse the police came again and enquired what I was doing. Once when up the Hungry Snout I caught a water bailiff who must have thought I was catching trout.

I think I was a born naturalist but in my late teens I had a second birth as I became convinced I was a sinner in need of salvation. This was a result of my father speaking to me. He told me that if I was to be saved I should accept Christ as my personal Saviour so I went upstairs that night and knelt down at my bedside and trusted Christ, it was as simple as that. My father's people had belonged to the Church of England but when he was in Blackpool my father was converted at Adelaide St. Wesleyan Methodist Sunday School where two women were holding a mission. My father came to see that baptism always followed faith in Jesus in the New Testament so he accepted believer's baptism by immersion so becoming a Baptist.

Once when I was on holiday at a farm near Lancaster the farmer's wife asked me to go to Scotforth with some shoes to be repaired. I found that the cobbler was a believer (Pentecostal) and we got talking about the baptism in the Holy Spirit. He said one should be baptized in water first. So I read right through the Acts of the Apostles and saw he was correct. They believed and were then baptized. I then told my father I wanted to be baptized as a witness to my faith in Christ my Saviour. This he did at Inskip Baptist Church. Later I contracted the habit of smoking, you know what lads are. First we smoked Dr Blossom's herbal cigarettes then a friend who had moved to New Moston near Manchester came and stayed overnight to attend a Christmas party at school and he gave me a packet of ten Players cigarettes. But my father had said that to spend money on cigarettes for a Christian was not good stewardship so a battle raged in my mind. However, I resolved it by throwing the packet over the Todmorden vicarage fence and I have never smoked again except in my dreams. For this deliverence I am very thankful. After all you might as well throw pound notes on the fire and let the smoke go up the chimney.

One of the things that much occupied my mind as a naturalist and a young Christian believer was that of evolution and creation. I read through Darwin's book on *The Origin of Species by means of Natural Selection* and could not see any contradiction to the evidence he put forward. I also saw that it supplied an answer to the problem of the presence of a lot of strange organisms such as *Plasmodium* the malaria parasite and a bird like the cuckoo which lays its eggs in other birds' nests. It therefore seemed to me that the Creator, being sovereign, had made choice of this method of slow unfolding because in His wisdom He saw it was the most efficient. By allowing freedom, however, there was also the possibility that things could go wrong, so that evil sprang out of the good, but the good had come first. To redress this possibility something more therefore was needed and this responsibility He took Himself by coming into the world in the person of Jesus His Only Begotten Son who completed the work of redemption by His death and Resurrection. His words on the Cross were "It is finished". Like my father wrote in his diary: "All our debts are wiped out". So now it is simply, "Be not afraid, only believe". That is the Good News in a nutshell. Such faith does not contradict science or knowledge but goes beyond it and the proof of the pudding is in the eating of it. Paul said, "Prove [test] all things, hold fast that which is good." So faith and science go hand in hand, and are not in opposition, indeed without faith science could not proceed. Every scientist must exercise faith or he would go nowhere and find out nothing. The atheist jumps the gun and says God does not exist. In this he may be true because God is not a created object but is outside space and time. "I am, that I am," says the Lord. "Men ought always to pray, and not to faint," said Jesus. That is the way forward. He still answers prayer but not always in the affirmative.

When I was only a boy of 14 and was shot in my left foot I had to learn to walk again and was on crutches for eighteen months. One day I went for a walk with two other boys up near Wickenberry Clough in Todmorden and we scrambled up a steep bank amongst bracken and rough grass. Then I discovered that I had lost the rubber shoe on the end of the left crutch so it sank into the soft ground and was almost useless. We searched and we searched but could not find it. My two pals finally got tired of this and left me to search alone. But I was not alone; I got down on my knees among the bracken and prayed for help. God answered my prayer in the affirmative because when I got up I walked straight to the thing and put it on the end of my crutch then I went and caught my pals up near Rive rocks and told them what I had done. So there is a Friend that sticketh closer than a brother. Paul had a thorn in the flesh and asked the Lord three times to take it away. But the answer he received was, "My grace is sufficient for thee, for my strength is made perfect in weakness." Some think it was his eyesight as he usually dictated his letters to an amanuensis. In his letter to the Galatians he mentioned the large letters he had written. Jesus also prayed in the garden of Gethsemane three times for the cup to pass from him. But it was not possible if you and I were to be saved. So he endured the Cross and thereby overcame the power of sin and death.

CHAPTER 2

Reminiscences of a Palaeobotanist in Lancashire and Yorkshire: 1935–39

I FIRST became interested in fossil plants as a student at Manchester University (1934–1939) where I took my Bachelor of Science degree in the Honours School of Botany (1937). I travelled daily to Manchester from Todmorden leaving by train at 7.45 am and well remember that the return journey cost one and tuppence ha'penny at the reduced "workman's" rate.

My home was at Ferney Lee Road near the Grammar School then known as the Todmorden Secondary School. Some members of the Todmorden Natural History Society attended evening classes held (in winter) at the Secondary School and organized by the Workers' Educational Association. The lecturer was Dr R.G.S. Hudson who later became Professor of Geology at Leeds University (Stubblefield 1966). Some of the interest aroused by these classes was conveyed to me although I did not attend the lectures.

Todmorden (SD*935242*) lies just within the West Riding of Yorkshire in three convergent valleys from Burnley, Rochdale and Halifax. My explorations in pursuit of natural history subjects, however, often led me over the county boundary into my native Lancashire.

The geology of Todmorden is that of the Millstone Grits which in the geological succession lie below the Lower Coal Measures of the Upper Carboniferous Period. These grits outcrop on the steep sides of the three principal valleys and in numerous delightful tributary cloughs. The principal river is the Calder rising between Burnley and Todmorden in the Cliviger gorge (SD*87–28-*). Over the back of the Pennines the coal-measures are "in the air" *i.e.* they are absent as a result of denudation but identical coal-seams outcrop on both the eastern flank in Yorkshire, and the western flank in Lancashire. The main grits thus brought to the surface in the Todmorden area are (in descending order):

Manchester University, the entrance I used, on Oxford Road near the Museum where there was a huge *Stigmaria ficoides*, showing to what a gigantic size the arborescent Carboniferous Lycopods grew, as described in E. W. Binney's monograph.
Photograph by courtesy of Dennis Print and Publishing from an original by V L Price.

1. The Woodhead Hill Rock.
2. The Rough Rock.
3. The Haslingden Flags.
4. The Holcombe Brook Grits.
5. The Hazel Greave Grit.
6. The Gorpley Grit.
7. The Upper and Lower Kinderscout Grits.
8. The Todmorden Grit.

Between the grits are shales and about twenty marine bands containing fossil shells of which the Goniatites are the most important as zonal index fossils by which the strata in different localities can be correlated. Goniatites were Cephalopoda related to the cuttle-fish and evolved from forms similar to the modern *Nautilus*. They were in turn the ancestors of the well-known Ammonites present in later (Mesozoic) rocks. The shells were chambered and as the cavities were filled with gas they gave the animals buoyancy. At Todmorden the lowest marine band was in the Sabden Shales exposed in Pennant and Oakhill Cloughs (SD927256 and SD938250).

During the year 1935 I made brief entries in my journal relating to my incipient interest in the geology of the area in which I lived and I can do no better than quote some of these jottings to show how this interest developed.

October 26, 1935. Did some geology up Oakhill Clough – bottom part.

November 2, 1935. Went up Oakhill Clough, found *Calamites*, Goniatites, and a little flat shell.

November 9, 1935. Went up Green's Clough – found *Sigillaria*.

December 12, 1935. Professor Lang gave us his last lecture of the session on Psilophytales.

December 21, 1935. Arranged to go up Dulesgate with A. D. Greenwood.

The latter was a fellow student who graduated at the same time as myself in Botany. He then lived at Heywood, Lancashire and had been stimulated to take an interest in fossil plants partly by a local older amateur naturalist. It was through A. D. Greenwood I learned of the remarkable petrifactions of coal-measure plants found in coal-balls, some of which had been obtained at Dulesgate between Todmorden and Bacup. Among the earliest students of the coal-ball flora was E. W. Binney (1812–81) a Manchester solicitor, who published his earliest work on the subject in 1855 in a joint paper with J. D. Hooker, on some specimens of the seed *Trigonocarpon*. Binney carried out further research and thereby laid the foundations for the later and more extensive work of W. C. Williamson (1816–95) which is embodied for the main part in nineteen memoirs published between 1871 and 1893 in the *Philosophical Transactions of the Royal Society*.

The chief successor to Williamson in this field of botany was D. H. Scott F.R.S. (1854–1934) who commenced his palaeobotanical work in collaboration with Williamson and at his suggestion. From its commencement the work on the coal-ball flora was carried out by means of ground sections but an important advance was made by John Walton while he was a lecturer at Manchester University, through the development of the peel technique for making serial sections of petrified plants (Walton 1928). This technique was later improved by the use of cellulose acetate film (instead of a solution) applied to the smoothed and etched surface of the coal-ball after flooding the dried surface with acetone (Joy, Willis and Lacey, 1956). These methods enable one to make about twenty serial sections from one millimetre thickness of material so that a detailed study of small plant structures such as seeds or sporangia can be made. It was A. D. Greenwood who first drew my attention to Walton's method of making sections of petrified plants by means of film pulls and in 1936 we both experimented with the technique after we had obtained some coal-balls. The following extracts from my journal show how this developed.

> *February 25, 1936*. Went up Dulesgate with A. D. Greenwood. Found our first coal-ball at Cloughfoot (SD*908238*) on the old shale tips. Continued up the valley on the top, the sun was shining brilliantly and there was snow on the ground. It was towards the end of the afternoon when we went down Green's Clough; found probable remnants of coal-balls. Caught 'bus at bottom.
>
> *March 19, 1936*. Today we had our exams (Botany-Physiology, and Physical Chemistry). At night Hamshaw Thomas gave a lecture on the origin of flowers. Professor Lang introduced A. D. Greenwood and myself to him.
>
> *March 25, 1936*. Went up to Manchester again. Spent all the day with Norman Holgate in the geology department grinding a coal-ball. Mr Patterson, the Geology Steward, took us into the basement and showed us a large coal-ball that Walton had taken

peels off. Made arrangements with A. D. Greenwood to go over to Heywood.

March 28, 1936. Went up Dulesgate on the 2.45 p.m. 'bus – a beautiful afternoon, Curlews, Mistle Thrushes, Meadow Pipits, Skylarks and Lapwings could all be heard. Found a beautiful little *Lepidostrobus*. Came home ladened with coal-balls.

April 4, 1936. A. D. Greenwood came with a friend of his on a motor cycle to collect coal-balls. I met them in the afternoon at Cloughfoot. We did not find anything exceptional.

April 22, 1936. Have just been reading parts of Hugh Miller's "Old Red Sandstone" and the botanical part of Tonge's "Coal". I have now made about 25 peels of coal-balls and have got some quite interesting things.

My technique at this time was to grind a coal-ball down from the surface as I had no cutting facilities. It was in this way and about this time that I discovered a specimen of the seed *Lagenostoma ovoides*. Fortunately I ground into its basal end and so did not lose anything important before I recognized it. Inside the megaspore was a well preserved prothallus bearing three archegonia at the apex. It was some time before I got another similar specimen and only after much searching. Another discovery in the Dulesgate nodules was the first specimen of *Botryopteris hirsuta* in which it became apparent that the stem was actually borne on a frond near the junction of a pinna with the rachis which bore it. Both these discoveries were later published (Long, 1943, 1944).

Having found coal-balls on the old tips at Dulesgate I began to search for them further afield as shown by the following entries in my journal.

August 8, 1936. Another lovely day; I walked up to Cloughfoot then got a 'bus to Bacup (SD868230). Went to Old Meadows Colliery (SD868238). Saw one of the workmen and got a promise of some coal-balls. Then went, after dinner, to Deerplay Colliery (SD870266) between Bacup and Burnley and found some on the tips.

After taking my final examinations in June 1937 I resumed the collecting of coal-balls as I was now seriously considering doing a year of research on them for an M.Sc. degree. I realized the importance of obtaining the material before the winter and this spurred me on as evinced by the following entries in my journal.

July 17, 1937. I have been to Meadows Pit in Bacup this morning and arranged to go over on Thursday during the first week of August. I shall probably be able to go down the mine. Apparently Duerden who was at Birkbeck College obtained material here. In the afternoon I went up Deerplay and struck down left to Nabb Colliery (SD*852261*). I asked a gentleman the way and told him I was after fossils. He pointed out the way but strongly recommended me to go to Heb Clough (SD*845254*) near Forest Holme south of Water. Here he said there was a colliery working and there were many fossils in the clough side. I shall have to go some time. I kept to my intended plan despite his strongly urging me to go to Heb Clough, and I was soon at Nabb. At one point on the right hand side of the rail track for the tubs, and not far from the top, a delph-like hollow was being filled in. They were putting huge coal-balls (average size about that of a rugby football) two deep and then covering them with soil and huge chunks of shale. Nearby was an older tipping place on which were many well-weathered coal-balls of smaller size than the others. I brought away half of a large coal-ball split in two very neatly.

I have been puzzling how to get the material home. The first difficulty is that the large coal-balls they were burying would be too large for me to work on, even if I got them home. The second difficulty is to get any of them (even smaller ones) home in a cheap way. No doubt the best way would be to get someone with a car or motor cycle with side-car to run me over. The only person I can think of is Mr Rushworth (a teacher at the Secondary School).

August 2, 1937. On Saturday morning I went on the 8.20 a.m. 'bus to Bacup. The morning was lovely. Bacup was very quiet – the end of their holiday week. I walked up to Deerplay and turned down left and back to Grime Bridge coal-pit (SD*850250*), owned by Hargreaves. There was a small tip but no coal-balls. I went over to Nabb and brought three coal-balls away.

Professor Lang wrote on Saturday and asked me to arrange my research problems in order of preference. I can only think of two practicable ones at present *viz* (i) investigations of the coal-ball flora; (ii) investigation of branching and vegetative propagation by root buds in Adder's Tongue fern (*Ophioglossum vulgatum*).

August 3, 1937. Another glorious day. In the morning I purchased one sack for 4d from Lewis Marshall's grocery shop. In the afternoon I went over to Nabb Colliery near Bacup with Mr and Mrs Rushworth in their car for the sake of obtaining fossils. We went via Dulesgate and came back via Towneley (SD *85–30-*). There is only a half mile difference either way. Coming back I noticed some sort of coal-tip near the brick-works (SD *858283*) above Towneley on the Bacup Road (A671).

The under-manager at Nabb was very good and offered to run one or two tubs of coal-balls down to the bottom of the track any time if I just dropped a line from Todmorden.

We acquired a heavy hammer from a nearby farmer (Luther King). It was very useful – actually indispensable for cracking the coal-balls. When cracked they are very deceptive and may look to have nothing in, yet when etched they may show quite a number of things. They seem to contain chiefly *Lepidodendron vasculare*; the large flattened stems invariably seem to lack the stele as though it had fallen out. One specimen showed what is probably such an isolated stele. The proof will be to see if there is any primary wood within the secondary wood in fair quantity. This would distinguish it from a *Stigmaria* stele.

What I shall have to do is, go over and crack the big specimens and bring halves away if cracked neatly. I should be able to bring about three decent sized halves home for 1/1½ d. The alternative would be to go by bicycle – this would involve a lot of walking.

September 6, 1937. On Thursday last I was interviewed by Professor Lang and he seemed to favour the idea of my doing research on coal-balls.

He suggested I should look over the specimens I have and number them noting anything worth working upon further. This I have done in part.

On Friday and Saturday I went over to Nabb Colliery. Both were good days. On Friday I discovered that many of the coal-balls which had been tipped had been buried. I therefore came away a bit disappointed but on Saturday I was able to dig them out and brought a number home.

September 7, 1937. In the afternoon I went over to Bacup (1.45 'bus) and walked over Deerplay to Nabb Colliery. I split quite a number of coal-balls and etched them. One large coal-ball contained *Lepidophloios fuliginosus*. I also got a *Lagenostoma ovoides* seed but without prothallus. The early part of the afternoon was beautiful but the sky clouded over and a strong wind arose;

coming back I caught the 'bus at Deerplay Inn (SD *866265*). When approaching Nabb Colliery I caught a partridge near an empty farm. It was unable to fly but did not seem to be wounded.

September 16, 1937. On Thursday last I went over to Bacup and dug some coal-balls out of the ground where they had been tipped at Nabb Colliery. On the way over I again saw the flock of partridges which I have seen a number of times now. The workman who tips the material at the pit was formerly one of the Brethren in Nelson. He is exceedingly interesting and intelligent to talk to though he is deaf. I gave him a Gospel of St John. At present numerous compressions in shale are being tipped including *Sigillaria*, Pteridosperm fronds, *Asterophyllites*, *Cordaites* leaves and others.

September 20, 1937. Today I have been at Bacup collecting coal-balls again. I set off on the 9.45 'bus and came back on the 8.15 p.m. from Bacup. Although at first unsettled the wind being in the east, the day cleared and became quite good. Most of the time I spent cracking coal-balls and etching with hydrochloric acid. I found two good stems of *Lepidodendron vasculare* with secondary growth. I also got a coal-ball with a seed of *Lagenostoma ovoides* half exposed longitudinally at the surface. Another peculiar specimen has a stele like a *Stigmaria* but surrounded by tissue very unlike that of a *Stigmaria* and apparently with leaf-bases.

September 26, 1937. Yesterday (Saturday), I went over to Bacup on the 9.45 a.m. 'bus; it was a glorious morning and fine all day – a great change from Friday when I went up Scaitcliffe (SD *923250*) to gather *Ophioglossum*, the rain falling all the time. Visiting the Nabb Colliery again I got a number of coal-balls one of which has *Sphenophyllum* in it, the first I have obtained. I saw the manager and he promised to get two tubs out of the pit towards the end of the week. I think it would be best to go over about Thursday. I must also try and go over Littleborough way. The Geological Survey memoir on "The Geology of the Rossendale Anticline" (W. B. Wright and others, 1927) should give me all the likely localities.

September 28, 1937. Yesterday I was at my father's shop all day in order to earn a little money for geological maps. [He was a bookbinder and machine ruler at Ridge Bank near Todmorden Station (SD *935243*)].

October 3, 1937. On Saturday I went on the 'bus as far as the

Bacup Road at Towneley near Burnley. I then walked up the Bacup Road as far as the Dyneley Coal Pit (SD858289) and asked one of the men I met there if coal-balls were to be obtained. He said that only a few occurred in the coal, the seam being the Upper Mountain Mine. I then asked him if he knew Black Clough mine (SD865277) and he directed me there. Arriving there I found a large number of ironstone bullions with goniatites but only a few coal-balls. The papers of the mine said that the seam worked is the "Arley" (at the base of the Middle Coal Measures) which is rather strange. I must look the matter up in the "Geology of the Rossendale Anticline".

After Black Clough I visited Nabb and collected a few coal-balls and then set out for Deerplay catching the 7.30 p.m. 'bus down to Bacup.

October 4, 1937. In the afternoon I went over to Bacup by the 1.45 'bus. I walked up to Deerplay and in conversation with a lady I learned that one can approach Deerplay from Holme Chapel. I should like to do this walk sometime.

At Nabb I obtained a *Lepidostrobus* but nothing of obvious value. The man who tips the waste shale etc, and who was formerly connected with the Brethren at Nelson, showed me a coin he had found when removing the surface soil to make more room for tipping. It said on Britannia and was dated 1587, – the time of Queen Elizabeth.

October 6, 1937. Yesterday I went up to Manchester and registered. I looked at some papers of Binney's in the Central Reference Library (published in the Manchester Lit. and Phil. Soc. Proceedings). They were not too helpful. Today I have ordered Sheet 76 of the Geological Survey (New Series) at Sherratt and Hughes.

October 13, 1937. Today I caught the 8.15 a.m. 'bus to Walsden Post Office (SD934220) and walked up Ramsden Clough (SD930212) bearing to the right. I visited the old mine right up on the moor above the reservoir and above the old Roman road (SD907216). This mine probably worked the Upper Mountain seam and no doubt was near the top Temperley's mine (SD897224) which I believe is called the Moorcock. I then came back and crossed the stream near the reservoir and climbed over the hill into Wardle (SD913166). Just as one tops the hill a large number of tips come into view above the new reservoir they are making. I searched many of these tips for coal-balls but never saw the faintest sign of their presence. In Wardle I asked a smith at his forge if there were any mines now working. He said there is none working now, the last being forced to close down when the reservoir started to be built. They seemed also pretty well worked out. I must go over to Wardle again and search between Wardle and Whitworth (SD*885195*) since on p.59 of "The Geology of the Rossendale Anticline" it is recorded that the Bullion Mine is still being worked. The man I asked said that the seams that had been worked were the Little Mine (Lower Foot) and 40 Yards Mine (Upper Mountain). The sequence of the seams is as follows (see Wright, 1927, p.31):

Name	Synonyms
Pasture Mine	
Cannel Mine	Upper Mountain in Rochdale area.
Upper Mountain	40 Yards; Yard Mine in Blackburn and Darwen; Top Bed in Burnley.
Inch Mine	Middle Mountain Mine; Fireclay Coal.
Upper Foot	Bullion Mine; Little Coal in Darwen; Bin Coal in Turton.
Lower Mountain	Ganister Mine; Half Yard Mine in Blackburn and Darwen, Yard Mine in Bacup.
Lower Foot Mine	Little Mine in Littleborough.
Bassy Mine	Salts at New Mills; Lower or Dirty Yard Mine at Rochdale; Shale Bed in Rushton.
Six Inch Mine	First Coal.
Sand Rock Mine	Featheredge Coal in Millstone Grit

From Wardle I continued into Littleborough (SD*939164*) and soon found myself at Shore (SD*929170*) not knowing where I was. Just past the Cottage Homes (SD*921169*) I looked over some old tips on the right hand side of the path approaching Littleborough but found nothing. Soon after, up to the left I noticed some more tips (SD*921173*), but being tired I decided to pass on and not bother to search them; actually this was the famous Shore Mine (reopened by Scott, Oliver, Lomax and Sutcliffe in the first decade of this century). I continued on my way and then saw some red-brick mills called Shore Mills. Knowing then that I was at Shore I retraced my steps and searched the aforementioned tips near Higher Shore Farm (SD*923172*). Almost the first glance showed remains of nodules and about the first one I looked at showed a petiole of *Anachoropteris*.

October 20, 1937. Today I have been in Littleborough, Wardle and Syke (SD*898158*). I caught the Littleborough 'bus at 8.55 a.m.

and then walked over to Syke, dipping down at Clough Bottom (SD*905152*). Between Wardle and Clough Bottom I found some *Ophioglossum vulgatum* growing in a field sloping down on the left bank of a stream, – the first valley one comes to after leaving Wardle. I searched the tips above and around Clough Bottom but saw nothing except bullions with goniatites in. I finished up by going round Hamer's Pasture Reservoir (SD*894162*) and dipping down Syke Road until on the Halifax road where I got a 'bus and was home by 3.15.

October 22, 1937. Yesterday was quite an eventful day. Norman Holgate had kindly seen Mr. Colvin at Burnley and the latter has offered to find out the mines in which coal-balls can be obtained. Norman also saw Mr Clegg of Bank Hall Colliery, Colne Road, Burnley (SD*845335*), where coal-balls are being brought to the surface. It would be as well I think to visit this mine next week if possible. The way to find it is to continue along Colne Road, across the Leeds-Liverpool Canal (SD*842333*) and look out for some big gates and a little office at the gates – ask here or further in for Mr Clegg.

October 24, 1937. Yesterday (Saturday), it suddenly became fine about 2.45 p.m. I decided quickly to go over to Bacup and caught the 3.15 'bus to Deerplay arriving at Nabb about 4 o'clock. I cracked about 30 coal-balls and etched them. I could not see properly so brought a number back and examined them at home when I discovered nothing of any worth. I learned that coal is being obtained in connection with the brickworks down Todmorden Road between Bacup and Sharneyford (SD*884244*); it seems as if it will be as well to go sometime. This I may be able to do next Saturday when I hope to go over to the opening of the Bacup Natural History Society Museum.

October 27, 1937. Today I have been in Burnley. I caught the 8 a.m. 'bus and soon found Bank Hall Colliery. The Manager, Mr Clegg, directed a man to show me where the coal-balls were. It rather looked as though only the smaller ones were sent up, being mixed up with the coal. After washing they got thrown out on a piece of waste ground. Many were too pyritous to be of much use but others were not too bad. I borrowed a large hammer and cracked twenty or thirty. *Lepidodendron vasculare, Lepidocarpon, Botryopterys hirsuta* and *B. ramosa, Bothrodendron* and *Calamites* seemed to be the main things present.

The colliery is quite a big one. There are three or four shafts and the following seams are worked; King Mine, Pasture Mine, Arley Mine, Mountain Mine (Union Mine). I feel certain that large numbers of other coal-balls, many much bigger than the ones I saw, must be produced and I suspect that they will be left underground.

I came back on the 12.45 'bus and was in Cornholme (SD905263) by 1.15. I left my case at Mr Crabtree's (the Post Office) and then went up Coal Clough (SD904272) with the hopes of finding *Ophioglossum* in the neighbouring pastures. Here I ate my dinner. I came down the next gully (SD907273) between Coal Clough and Shore (SD911267). No stream, however, was running in it for some reason or other. The bed had an old wooden channel. Down the right side of this gully is a vertical exposure of shales. Bullions had fallen out of this into the bottom and one contained a plant stem, apparently a *Lepidodendron*. I managed to split a flake off the outside and it seemed to be in a halonial condition; the stele was present and one could just make out the xylem which seemed to be medullated.

October 31, 1937. On Saturday morning I went up Dulesgate and spent two hours on the tips so filling my rucksack with coal-balls. One has a fairly large stem of *Lepidophloios fuliginosus* in it. I intended walking over to Bacup but about 12 o'clock it became very dark and rained hard so I came back home and changed, then took the 2.15 'bus over to Bacup and went direct to the new rooms of the Natural History Society underneath Zion Chapel. A little museum was opened by Dr. Jackson of the Manchester Museum. It appears the former premises were opened by Professor Boyd Dawkins. I came back with Mr Ogden of the Todmorden Museum at Centre Vale. He said he thought he could get the Corporation car for bringing coal-balls over from Bacup.

November 5, 1937. Today I took the 8.20 a.m., 'bus to Bay Horse Inn (SD901236), Dulesgate. The fog was very bad here although there was scarcely any at Todmorden. I went to South Graine Colliery (SD893238) and enquired if any coal-balls were being produced. It does not seem that there are any. I have just been reading in "The Geology of the Rossendale Anticline" that the union of the Upper Foot and Lower Mountain mines was proved at South Graine. A carter who works at South Graine said that the chief miner would be up out of the mine about 11.30 for his dinner and he might be able to give me information. I left

South Graine and scrambled up the tips until I came to the air shaft and the road which leads round to South Graine Farm. Here I met Mr J. H. Temperley. He asked me if I was looking for something and I told him I was looking for coal-balls. He mentioned that they were getting either coal or fireclay (I forget which) from the Bassy Mine and that they were getting fireclay from beneath the Upper Foot seam. He also said that the Sand Rock seam was got just below Bay Horse Inn. All this is very puzzling because the Cloughfoot mine is so much lower down the valley. The reason they are not getting the Upper Foot coal is because it is flooded. It seems that the firm who worked it were Howards of Rochdale (or Littleborough) and they got it at the middle of its dip. This meant that one part could be drained but the other could not. Mr Temperley said it would need some powerful pumps to make its working possible or else to get it higher up on the hill (left-hand side of the road going down). This outcrop may be worth looking for.

Mr Temperley told me it might be worth my while to see Mr Thomas Ashworth, who is manager of Hargreaves Collieries and who lives near Broadclough, Bacup (SD 867240), some such place as Woodland View. Mr Temperley showed me a fault which might explain the difference in altitude of Moorcock Upper Mountain seam and South Graine.

I had a look on the tip associated with the old Sharneyford Colliery on the right-hand side of the road going down to Bacup. There is a tip also on the left: I did not find anything.

I next called at the brickworks on the left going down to Bacup; coal is being obtained here and has only been mined about 12 months. It seems to be the Lower Mountain and is a good three feet thick. I asked if it was the same seam that is got at Meadows and they said it was. It is on land belonging to Parrock Farm (SD 884238) which is roughly opposite Carlton Club. Here, on the left-hand side going down, just above the children's recreation ground is a clay pit. They are getting clay which is below a seam of coal about 12 to 18 inches thick. Formerly there were 2 or 3 mines here, just near a dam. Now they are clearing the surface soil off, exposing and removing the coal, and getting the rich yellow clay about 6 feet thick. This they make into bricks. I think it must be the clay below the Upper Mountain mine which means that the Upper Foot should outcrop somewhere between here and the brickworks. It may be worth looking for. The exposure is really beautiful and worth photographing sometime.

In the afternoon I went up to Deerplay and saw the under-manager who promised to get me one or two more tubs of coal-balls out. I came across the compression of a *Lepidodendron* in the *Ulodendron* condition. The scars were about the size of a florin.

November 6, 1937. At the invitation of Norman Holgate I have spent the day in Burnley. I went on the 9 a.m. 'bus and met him at the Wellington Hotel. We first walked along Brunshaw Road and had a look at Bee Hole Colliery (SD*850326*). This has now ceased working and the nature of its head gear suggested that it was not a very deep mine, probably working some seams of the Middle Coal Measures. We next caught a 'bus for Worsthorne (SD*875323*), 3d. fare, and were soon on the Gorple Road which runs very straight and soon gets out onto very rough bleak moorlands. Norman pointed out to me Hurstwood Reservoir and near here numerous grass-covered hillocks believed to be the remains of an ancient lime industry (drift lime of course). Nearby he pointed out to me an overflow channel. We left the footpath and followed the stream on the right striking down at about the first bench mark on a stone stoop on the right-hand side of the road. In this area the lower coal measures just begin. We found no indications of natural exposures in the stream because of the covering of drift. Further up we searched some bare patches in the moor. Here were numerous indications of dead birch trees, the altitude being 1150 feet.

Further up, the stream runs under the road. At this point we turned off the path towards the left and followed alongside a wall over Wether Edge (SD*903326*) to the next valley (Swindon Water). Here we searched for and found the Bassy Mine, on the right-hand side going down. It had been recently artificially exposed and consisted of 2–3 feet of coal and shaly layers.

Further down the map shows the presence of the Lower Mountain (Union) Mine and an associated Marine Band but this we could not find. Lower down still, below the sheep-fold, the Upper Mountain Mine is shown but we did not find it. We saw an example of current bedding in the stream.

Norman mentioned that coal-balls had been sent from Bank Hall to Yale University and they were highly delighted with them. He also suggested that I should get in touch with Dr W. B. Wright of the Geological Survey Office, Oxford Road, Manchester. This should be well worth while.

I should obtain the following quarter sheet geological maps (6 inches to the mile):

(i)   Bacup. Sheet 72 S.W.
(ii) Todmorden. Sheet 73 S.W.

Also the new series one-inch maps: Sheet 76 and Sheet 77.

November 11, 1937. In "The Geology of the Rossendale Anticline" p.148 it mentions that in the Rochdale area small workings of the Lower Foot, Upper Foot and Upper Mountain Mines occur.

November 13, 1937. I have been over to Nabb again today. I went on the 8.50 a.m. 'bus as this is the last one on which one can get "workman's" rate. It was very cold all the day, the sky being clear and the wind in the north. I first visited the old pit at Broadclough but there are no tips there; in a gully nearby there is a clay patch as if the seam outcrops here. I walked up to Deerplay and because of the north wind could distinctly hear the trains and general hum of town activity at Burnley. Coming back in the afternoon Pendle Hill (SD*804414*) was very clear.

The manager had forgotten to get me any more coal-balls out of the pit so I went over the old material and this was fortunate as I am almost certain I came upon a couple of stems of *Botryopteris cylindrica*. I saw the under-manager and he promised to get some coal-balls out for Monday or Tuesday. The next step is to see Mr Thomas Ashworth, 1 Woodland View, and arrange for two tubs to be got out from Meadows, from as near Sharneyford as possible.

I have been to a lecture tonight on "The Geology of Todmorden" by Dr Hudson of Leeds University – his treatment of the subject was fairly elementary. A collection of goniatite compressions was on show.

Mr Lawrence (a neighbour) mentioned to me that an old man named Rothwell, who formerly had a grocer's shop in Sowerby Bridge (SE*05–23-*), has some specimens from Dulesgate. Mr Lawrence said these were polished. I really must write to him and try and arrange to see him. Dr Hudson seemed to think that all the localities where coal-balls had been got in Yorkshire were now out-of-date and closed. He said that about five years ago, at an International Congress of Botanists in Leeds, some coal-balls were required but none could be got except a number kept over from the old collecting days.

November 17, 1937. Today I have been to the Geological Survey

Office in Manchester to see Dr W. B. Wright about coal-ball localities; he was most willing to converse on the matter but was not too informative; it appears that the survey of the Rossendale area was about the year 1921. He introduced me to Dr Wray. He also showed me two papers on the correlation of middle coal measures by means of microspores and suggested that it might be possible to give names to them. At present they are only known by numbers. He also mentioned Maslen as having done quite a lot at London (I think he said Kew, where D. H. Scott was). He mentioned that the Lower Coal Measures have been worked west of Wigan but that they were unable to correlate them. He said there was a marine band which possibly corresponded to that of the Upper Foot seam but that it was possibly not this. It rather looked as if they were preparing a memoir on the geology of Wigan.

The two papers by Raistrick of Armstrong College, which he mentioned in connection with the microspores of the coal-measures, were – (i) The North of England Institute of Mining and Mechanical Engineers Transactions 1932–3, Vol 83 and (ii) *ibid* 1934–5, Vol 85. He said that no such work had been done on the Lower Coal Measures. The technique briefly was:

(a) Grind coal to pass through a 200 mesh.
(b) Three days in Schulze's solution; decant off Bismark Brown.
(c) Strong alkali three days.
(d) Do counts of spores and make graphs.

He said that as one ascended through the measures the rarer spores became more common. This he interpreted as showing that an upland area from which the spores were being carried by wind was coming nearer. I question this.

Dr Wray referred me to a place near Newhey (SD*935116*) on the western end of Ogden Reservoir (SD*953124*) where he said the Bullion Mine was formerly worked – on the map there is a fault nearby. He also said that it outcrops south of Rakewood (SD*944142*) near Tunshill (SD*944134*). He also suggested I should ask at Haugh Hey Colliery (SD*958112*) as to whether there are any others in the vicinity. For getting to know about the Oldham district he said I could enquire at Dransfield's Starring Fireclay Works, Littleborough (SD*92–16-*), since they came from Oldham. He offered to lend me reprints of any papers and said he would look them out *eg* Stopes and Watson, 1908.

This morning Norman Holgate brought me a list of the Burnley collieries. They are as follows:

Hapton Valley (SD*810313*)
Calder, just short of Altham Church; approach from Padiham, (SD*775331*)
* Moorfields near Accrington, (SD*77–28*-)
* Rishton (SD*72–28*-)
* Scaitcliffe near Accrington, (SD*747277*)
* Broad Oak, (this may be a place near Halifax (SE*125246*), near Sunny Bank)
Towneley (SD*848307*)
Boggart Bridge (SD*853307*)
Cliviger (SD*884275*)
Bank Hall (SD*845335*)
Trawden (SD*912388*)

Those marked * are not the Union seam.

November 19, 1937. Received a post-card from Mr Rothwell of Sowerby Bridge. He suggests I go down any Wednesday afternoon. His address is 14 Myrtle Terrace, Tuel Lane, Sowerby Bridge, Yorkshire (SE*05–23*-).

November 24, 1937. This morning I went up Dulesgate on the 8.20 'bus booking workman, – 3d return to Cloughfoot. I gathered enough coal-balls to carry and was home by 10.45 a.m. It was a really beautiful morning, no fog, a clear blue sky, and slight hoar frost.

In the afternoon I went to Sowerby Bridge on the 1.17 p.m. train (one shilling and fourpence return) to see Mr R. Rothwell. He proved very interesting to talk to. He makes up 80 tomorrow. He had numerous fossil slides chiefly made by Mr W. Hemingway. Mr Rothwell, it appears, collected material and sent it to Hemingway who was then at Barnsley. He showed me a letter from Hemingway in which Binns of Halifax was mentioned. Hemingway mentioned how Binns always did his grinding down on the sinkstone. Already his landlord had put two new ones in and could not understand how it was they went so quickly. He had a beautiful preparation of *Stauropteris burntislandica* from Pettycur material, made by Hemingway.

Hemingway said he could make 18 sections per inch. Mr Rothwell had known Dr Kidston and had him at his house. The microscope Mr Rothwell possesses is a handsome Swift Binocular petrological microscope and belonged to Robert Law, having been presented to him on the occasion of his marriage. He had to do with the discovery of the urns near Cross Stone in Todmorden, (SD*943254*).

Mr Rothwell showed me a beautiful arrow head found in a runnel of water on Boulsworth (SD*929355*). He searched for four years before he found it though he came across many flint chippings. He confirmed the saying that they are only found on the sandy gravel underlying the peat.

Mr Rothwell knew Cash and James Spencer very well. He showed me his material in the cellar and gave me a few coal-ball fragments from Dulesgate and Sunny Bank, Halifax.

The following coal-ball localities were recommended by Mr Rothwell:

1. Ford Hill Pit, Queensbury (SE*10–30-*), between Halifax and Bradford.
2. Sunny Bank Pit (SE*117247*); go to Halifax North Bridge and ask for Southowram car to Bank Top, (SE*105244*).
3. Old workings at Queensbury Station, (SE*10–30-*).
4. North Deighton, (Hemingway got good material here (SE*38–51-*).
5. Booth Hill and Sunfield Pit, Oldham, (SE*38–51-*).
6. Fieldhouse Coal Pit, Bradley, near Huddersfield (se*95–07-*).
7. Parkfield Coal Pit, Oldham (?SD*929020*).
8. Towneley, near Burnley (Mr Hemingway said this was a good place).
9. Bullhouse, Yorkshire (SE*202030*).
10. Crompton Moor, Lancashire (SD*95–10-*).

November 27, 1937. I went over to Bacup on the 9.50 a.m. 'bus and also rode up to Deerplay Inn. I was soon at Nabb where the under-manager, who is called Howarth I believe, had arranged for another lot of coal-balls to be got out. I went through about two thirds of them but did not find anything special. The best one had a number of seeds of *Lagenostoma ovoides* in and *Lyginopteris*. I asked the farmer's wife the correct address of their house and she said, Nabb, Lumb in Rossendale, though she said they must actually come in Water. I filled my attaché case and got up on to the Deerplay Road when a man in his car drew up and offered me a lift. Actually he thought at first I was someone else. I was going to walk down to Broadclough so that it was a great saving. He put me off at Broadclough and I called on Mr Thomas

Ashworth, who proved to be a very nice gentleman and very obliging. He showed me much correspondence of Dr Duerden's at Birkbeck College, the last being just in the new year of 1935 so that he must be still working. Evidently Duerden cuts the coal-balls as he mentioned about his instrument being repaired on one occasion.

Mr Ashworth promised to get me some coal-balls out from as near to Sharneyford as possible and said he would drop me a post-card. He showed me plans of various mines and told me definitely that at both Grime Bridge and Hile the Upper Foot and Lower Mountain seams are not united and only the latter is worked. The union takes place in Nabb Colliery and he showed me how it is in the nature of quite an irregular line. He showed me a book on the geology of Lancashire and offered to lend it to me. It might be worth while to take "The Geology of Rossendale Anticline" sometime and the geological map since he is very interesting to talk to. He mentioned how Old Meadows Colliery works right up to the fault at Sharneyford which has faulted down all Dulesgate.

December 4, 1937. Professor Weiss, who came into the University last Saturday has not been well and is now 72. Professor Lang is, I believe, due to retire next year.

December 6, 1937. Mr W. N. Croft came in today for the first time.

December 18, 1937. An American palaeobotanist working under Hamshaw Thomas at Cambridge and originating at St Louis I think, called H. N. Andrews, has spent one or two days here looking over the fossil slides. He has recently been in Glasgow also. Professor Lang goes to Wales for Christmas I believe, from next Tuesday I think. I believe the course of this term's work will be the last he takes.

December 23, 1937. I looked up the position of Newhey on the 1 inch Ordnance Survey map in the Manchester Central Library. The colliery is on the west end of the first reservoir (Ogden Reservoir) and is called Rough Bank Colliery (SD947123). Haugh Hey Colliery (SD958112) is marked as disused. Tunshill Colliery (disused) lies N.W. of Rough Bank Colliery.

December 28, 1937. On Friday last I went to Rochdale on the 8.50 'bus and from thence to Newhey (this latter part of the journey cost 4½ d each way). Having arrived at Newhey I first went to the old Jubilee Colliery (SD944109) but did not find any tips. I then retraced my steps into Newhey and went up the Huddersfield Road. There were some tips on the right-hand side, but nothing on them. I eventually arrived at Rough Bank on the western end of Ogden Reservoir. Coal is still being worked in a small way. The Little Mine was being worked and a man said the Yard Mine also has been worked; these correspond to the Upper Mountain (Yard Mine) and the Lower Foot. I returned into Rochdale and got on a 'bus for Littleborough and went out to Shore. Here I collected what I could and dug a number of coal-balls out of the tip itself. There must be numbers still buried.

December 30, 1937. I have written to the Yorkshire Collieries and Fireclay Works which are most likely places for coal-balls. Last night I came across *Stauropteris* and *Calamostachys* in some of the Shore material.

January 8, 1938. This afternoon I walked over to Dulesgate via Dobroyd (SD930238). I enjoyed the walk, typical January, the year just at its opening. I found a *Sphenophyllum* and a *Lepidostrobus* amongst some coal-balls at Dulesgate. I accidentally dropped my pocket microscope near the stream and had set off without it but discovered my loss and went back and found it. I then caught a 'bus.

January 12, 1938. Yesterday and Monday I stayed at the University till 9 p.m. I am trying to make drawings of *Bothrodendron mundum*.

January 14, 1938. As I came across the Quadrangle at the University today I heard a Mistle Thrush singing on top of a stone pinnacle, very high up above the Natural History Theatre. It sang beautifully and brought back associations of wild hills and cloughs.

January 27, 1938. Yesterday I lectured at the Halifax Scientific and Philosophical Society, Harrison Road, on "The Fossil Flora of Coal-Balls".

February 5, 1938. This afternoon I went by 'bus to Littleborough and then walked out to Shore. By digging on the tips I got quite a number of coal-balls many of them embedded in clay. Others showed that they had been half in coal and half in clay. When I got home and examined them I found most of them not too good. About five had good *Stauropteris* in and one *Trigonocarpon*, others *Lyginopteris* but no *Bothrodendron mundum*.

February 13, 1938. Yesterday I went to Bacup on the 9.45 a.m. 'bus. I went to Old Meadows Colliery and managed to see Mr Hunter who promised to obtain for me some coal-balls for next Saturday. In the afternoon I went to Shore at Littleborough and dug up some more coal-balls. It has been a very cold week-end, the wind blowing strongly from the N.N.W. We have also had slight falls of snow. It was like a gale at Littleborough when I was digging.

February 16, 1938. Today I have been out collecting. I caught the 7.45 a.m. 'bus to Hebden Bridge (SD*993272*) and booked workman return 6d. I could not book return right through to Halifax. At Hebden Bridge I got a Halifax 'bus but could not book workman since it was after 8 o'clock. I got a 'bus for Queensbury (SD*10–30-*), and was soon out at Ford Hill. I found the manager Mr Brown a very nice fellow. He first took me over the tips where we found slight indications of pyritized coal-balls. He next suggested that I should go down to where they are working the Hard Bed. I first changed one boot (the non-surgical one) and put on an old coat. We had to wait for a shot being fired before we could go down. In going down, the wire ropes by which they signal got crossed and we had to descend slowly. Two seams are worked, the Hard and Soft beds.

The cage used for going down in is also used for hauling up the tubs. Both coal and clay are brought to the surface this way. In the actual workings it is possible to stand upright since both coal and clay are removed. The coal seam is about 22 inches thick underlain by about 6 inches of gannister and then about a yard of fireclays. The roof of the seam is shale with ironstone balls. *Pterinopectens* were abundant and I got a large goniatite. I found all the workmen very decent. The manager Mr Brown is leaving the pit tomorrow and taking over at a clay pit (Allen's), I believe, somewhere near Sunny Vale. He said that I had come at a very fortunate time, since, owing to the fact that he is leaving he could bother with me. The pit opened in 1930.

I left Ford Hill just before noon; being somewhat disappointed I was half inclined to go home but after visiting Woolworths I decided to go out to Lightcliffe (SE*135255*) between Hipperholme (SE*12–26-*) and Brighouse (SE*14–23-*) to Walter Clough Pit (SE*120246*). This actually is not far from the old Sunny Bank (SE*117247*) and Shibden Hall pits (SE*114254*). I again found the manager very obliging. He had got some coal-balls out specially for me, provided a hammer and a man cracked the coal-balls for me. They contained *Lepidodendron vasculare*, *Stigmaria* with large pith, *Sigillaria*, *Lepidostrobus*, *Lepidophloios*, *Botryopteris hirsuta* and *Lyginopteris*. The mine has been working continuously for 33 years. Before that it had worked and then closed down. Originally a water-wheel was used to work the pumps. The coal-balls were on the whole very similar to those from Nabb and Burnley. Mr Brown mentioned two other places where he knew the Halifax Hard Bed to be worked *viz* Crow Edge (SE*187045*), on the road to Penistone (SE*23–03*) from Queensbury or on the left-hand side of the road from Huddersfield to Sheffield (Stocksbridge SK*27–98*-); also Bull House (SE*212026*). Here clay and gannister are obtained. This is past Crow Edge and Flouch Inn (SK*197015*) where one turns left to Penistone.

February 26, 1938. I was up before 7 a.m. and caught the 8.20 'bus for Bacup and arrived at Old Meadows Colliery about 7 minutes to 9. Mr Hunter was in his little office. He ate his breakfast, did something (possibly tempering) with some cold-steel objects which he had in the fire, changed his clothes and then we were ready for going into the pit. In the meanwhile he commented on my being there to time. He had two candles fitted into long tin boxes which acted as reflectors. On the tin box was a leather handle. Next we went to the pit entrance, oiled the trolley wheels and were ready for starting. One held the lamp and the front of the trolley with the left hand; at the right-hand side was a brake in the form of a narrow board hinged to the edge of the trolley so that it could be allowed to rest on the wheels and pressed down. One knelt near the back of the trolley with the left leg and pushed with the right. The distance a collier goes each day is about a mile, some slightly uphill, some a bit downhill. It is hard work for anyone not used to it. Mr Hunter could not remember his first reaction to it since he started work at the age of 12.

As we went along the travelling road he pointed out to me the 6 inch mine. The tunnel was very low at parts but of a decent height (about 4 feet on average). It was of course impossible to stand upright except at certain places. There is great skill in pushing one of the trolleys along the lines since it means pushing off the sleepers each time. The ground between two sleepers consisted of wet mud, puddles, or running water. The men must get thigh muscles like iron. I had to stop and rest a number of times. The candles made wonderfully good lamps and ours lasted the whole 2 hours we were in. Mr Hunter pointed out how the coal surface has to be dusted over (by regulation) with a sort of lime dust, to lessen the risk of explosion. It turns a deep brown colour. In some places there was a lot of water and the rails are constantly wet and hence corrode. It is important that they should be looked after because one can go at quite a fast rate down the short declines. In fact this travelling at a quick rate is quite a hair-raising experience and if anything should happen one would be terribly thrown on one's face. One collier broke a collar bone in such a way. At one point Mr Hunter pointed out to me a "wash-out" in the coal, apparently caused by some ancient current of water. After we had gone about half a mile we came to the union where the Upper Foot seam joins the Lower Mountain seam, the latter we had been following for some time (it is almost horizontal). Here we could see numbers of coal-balls in situ. We left our trolleys and went on a bit further past two places where a piece of rather stiff sacking was hung across the rail-road. These, Mr Hunter explained, were for the purpose of directing air currents. Next we retraced our steps to our trolleys and turned down a side road and crawled a long way on our hands and knees. Mr Hunter carried the sack and I picked out what coal-balls I wanted. These he put on his trolley and kindly took out for me in this way. The only men we met who were working were making channels and drains for the water and putting new rails down.

March 12, 1938. This morning I went over to Bacup (9.45 a.m.). Mr Hunter at Old Meadows had got me some more coal-balls. They were a very nice size but contained very little of interest. He obtained them from 50 yards beyond the union in the double seam. He showed me the place on the plan of the mine. The first lot we got (a week ago) were from the actual place of union. These latter contained *Calamites, Botryopteris cylindrica, Ankyropteris corrugata* and *Botryopteris ramosa*, whereas the second lot (from beyond the union) only contained *Lepidodendron vasculare, Botryopteris ramosa* and *Stigmaria*. I think the best policy will be to give Old Meadows a rest for a week or so and then ask for some to be got out from near the union towards Sharneyford direction.

March 16, 1938. I caught the 8 o'clock 'bus to Burnley and got straight on an Accrington 'bus for Hapton Valley (SD810313). I found the manager a big-built man eating his breakfast in his office. It appears they only just touch the union of the two seams and work chiefly the Lower Mountain Mine, the reason being that the Upper Foot is so sulphurous. They get the coal chiefly for coke making and must keep the sulphur content down since it would depreciate the quality of the steel made with the help of the coke. He took me to the tips where about four men were tipping shale etc and about twelve to fifteen out-of-works were picking coal. The manager cleared them off making one man leave his half-full sack. He contended for it saying the sack was his. The manager had it sent away. I saw the man later, however, and he said he got it back.

In the afternoon I went to Towneley. The manager is a young fellow. He was very decent and had got me some coal-balls in a sack. These were taken into the place where a number of forges and a steam-hammer were. A man cracked the coal-balls for me and I etched them. The plants present were *Botryopteris, Calamites, Lepidodendron vasculare, Lepidophloios fuliginosus, Lyginopteris, Sigillaria* and *Stigmaria*. They were, however, badly pyritized.

March 30, 1938. I have been in Burnley all day, at the Towneley Colliery. I went on the 8.30 a.m. 'bus. I learned that the Mountain Mine, although formed of two seams, is a perfect union in that the two seams are not separated in any way by shale, although sometimes there is a layer of pyrites. I also learned that Boggart Bridge Mine is the same Company. In fact the coal-balls which were sent out were from that pit.

I proved the presence of *Botryopteris cylindrica*, also *Sigillaria* (probably *mamillaris*). I did not come across any *Bothrodendron mundum* this time although there was a fair quantity in the first lot of material. *Lyginopteris* occurred commonly and *Botryopteris* but nothing more.

April 9, 1938. This morning I travelled up to Manchester by the 7.45 train and was at the University all morning. At 12 o'clock I went out with D. Doxey and had lunch at his home. I met Mr and Mrs Doxey for the second time. Soon after 2 o'clock we set out by car for Crow Edge. We went via Denton (SJ*92–95*-), Hyde (SJ*92–94*-), Tintwhistle (SK*02–97*-), and the Longdendale valley (SK*0–9*-), climbing to 1400 feet. The day was on the cold side with a N.E. wind. We passed the Woodhead railway tunnel (SK*09–99*-) which is 3 miles in length and turned up the Huddersfield Road at the Flouch Inn (SE*197015*). We found Crow Edge (SE*187045*) all right and Mr Smith Howard who had worked there 60 years, and got onto the tips. There were dozens of coal-balls but all badly pyritized. One had a *Lepidophloios* halonial branch, others showed *Lyginopteris* and *Lepidodendron vasculare*. We came back via Carlecotes (SE*17–03*-) and Dundford Bridge (SE*16–02*-) where there is a watershed, the one valley draining into the Don (SK*17–99*-) and so to the North Sea, the other valley containing the Derwent (SK*13–97*-) which eventually runs into the Irish Sea. Mr Smith Howard was a typical West Yorkshireman, big, with a gold watch chain, black suit, slow moving but all there, with a good head on him; he filled his pipe twice from Mr Doxey's pouch but would not accept anything else. He said that eight men only worked at Crow Edge Colliery and that all the coal went to one mill. A nearby colliery worked the Better Bed for coking purposes. We got back soon after 7 o'clock and took the coal-balls to the University.

April 13, 1938. Yesterday afternoon I went out to Moorside, Oldham (SD*953073*); it cost 1/5d in 'bus fares. I visited three old pits, the first near Moorside Church. It had not worked during the last 50 years. When it did, the men received bread and eggs for payment, not money. The third place I went to was Booth Hill. Here I got several coal-balls of the type identical with Shore material. In them was *Lyginopteris, Lagenostoma, Stauropteris, Calamites, Lepidostrobus, Sphenophyllum* and *Stigmaria*.

April 28, 1938. On Saturday last I went over to Laneshaw Bridge (SD*922406*) near Colne and found the old coal pit and obtained some material showing at least *Lyginopteris*. I collected some Pectens and Goniatites also and took them to Dr Jackson in the Manchester Museum on Tuesday. He could not say definitely, however, whether they were from Lower Coal Measures or the top of the Millstone grits.

Today I have been in Bacup all day *viz* at Deerplay Colliery. I went through a fair quantity of material and found *Lyginopteris, Lagenostoma, Lepidophloios fuliginosus, Lepidodendron vasculare, Sigillaria, Bothrodendron mundum, Botryopteris hirsuta, Stigmaria ficoides* and *S. bacupensis*.

I walked up to Deerplay from Old Meadows by way of the road past Irwell Springs. In a little clough (SD*873251*) just below the day-eye near the top air-shaft to Old Meadows Colliery I found a coal-seam about 9 inches thick and underlain by 2 feet of fireclay. It was probably the Lower Foot seam.

April 30, 1938. In the morning I went to Old Meadows Colliery, Bacup, by the 8.15 'bus and came back at 11.15 a.m. I obtained a large number of petioles (about 13) of *Botryopteris hirsuta* in one coal-ball, also *Sphenophyllum plurifoliatum*.

In the afternoon I went on a ramble with the Todmorden Natural History Society, leader Mr Sutcliffe, to Ramsden and Whiteslack Cloughs (SD92–20-). We commenced with an exposure of Kinderscout Grit and went up Ramsden Clough past the dam and along the path to where it crosses the stream. Here we saw the Gorpley Grit. Then we retraced our steps and went up Whiteslack Clough. Here was pointed out the Hazel Greave Grit up on the left at the edge of the valley, the underlying Gorpley Grit being covered by landslips. Very near the stream bed just by a little square sort of dam (perhaps 4 yards × 4 yards) and at the stream side we found a marine band, very hard to get at. Mr Rushworth and I worked for 20 minutes to get the goniatites. We were thus left behind and went up the rest of the stream very quickly. We then turned off right and caught up with the party but while going up this little side stream we came across a shaly coal on a fire-clay, below the Holcombe Brook Grit. At the top of the waterfall above the Holcombe Brook Grit was a marine band with *Pterinopecten*, *Posidonomya* and *Gastrioceros sigma*. Here Mr Rushworth and I were left behind again and went over the moor and down into Ramsden again.

June 1, 1938. Tomorrow Dr Stockmans returns to Brussels. He has just had a fortnight here.

June 2, 1938. Today I purchased two six-inch geological maps, Todmorden and Shore Moor.

June 8, 1938. Yesterday turned out a glorious day. I was out all afternoon and evening. I caught the 1.15 'bus to Walsden Waggon (SD93–21-) and went up Ramsden Clough (SD92–21-). Just above Ragby Bridge (SD923215), on the right-hand side going up, I saw a Small Copper butterfly and immediately after a Small Heath which I caught. Soon after I saw another. I proceeded up the clough and then turned right and got onto the Foul Clough Road (SD922219), continuing past Lower and Higher Ditches (SD916214) and Moorcock (SD905212). I examined the 6 inch mine above the Rough Rock in an old quarry at the top end of Ramsden Reservoir (SD912210). Up Foul Clough (SD910209) I found a thick coal seam near to a mine mouth under a water-fall. This I took to be the Lower Mountain Mine but it may have been the Upper Foot. It seemed to be a yard thick. I went across Deacon Pasture where there is an old mine mouth

(SD907209) near which I found the Bullion Mine in a little trickle of water. It seemed very pyritous. This is very near to the Foul Clough and on the right. Across the moor there are five shale heaps, all Upper Mountain going up. I went over to the Foul Clough Colliery (SD907205) but found nothing there, then round the edge of the hill to Wreck Beds (SD904223) where I found the Lower Mountain, Bullion Mine and Inch Mine. In another gully further on was a thinner 1 foot seam, probably the Lower Foot. There were many Curlews and a pair of Golden Plovers. I turned down Howroyd Clough (SD912231) and watched three persons rowing a boat on Gorpley Reservoir (SD910230). Two Common Sandpipers were flying over the water. After collecting a number of coal-balls at Cloughfoot I caught the 9.30 p.m. 'bus from Cloughfoot and was home by about 10 o'clock.

June 15, 1938. On Monday I received some literature on coal-ball plants from Dr Stockmans and Dr Koopmans.

June 25, 1938. Yesterday an announcement placed on the notice board of the University stated that I have been awarded half of the Grisedale Research Scholarship along with Joan Raby. This means £100 for each of us.

August 3, 1938. Yesterday I went to Walter Clough Colliery at Halifax and examined coal-balls. I found *Lepidodendron vasculare*, *Lyginopteris oldhamia*, *Lagenostoma ovoides*, *Lepidophloios fuliginosus*, *Botryopteris hirsuta* and *B. cylindrica*, *Sigillaria*, *Stigmaria* and *Ankyropteris corrugata*.

August 16, 1938. On Saturday last I went up Dulesgate walking over by way of Dobroyd. Great havoc has been done by a terrific thunder storm which occurred last Friday morning. I was in Manchester Central Reference Library at the time. When I left home it was raining and the valley was full of dense mist the whole air being absolutely saturated, moreover it had been thundering from 4 a.m. to 7 a.m. Two houses were struck near my Father's shop and Mr J. Barker's on Garden Street. The water swept off the Sourhall side of the valley (SD91–24-) down into Ewood Lane (SD927248) making new channels and bringing boulders and dirt down, covering the lawn at Wood Cottage. In Dulesgate the road was blocked at two places and up Walsden the railway was blocked at 3 or 4 parts, also at Dobroyd where the Old Rake was carried away. Returning from Manchester we had to get out of the train at Littleborough (SD939164) and come home by 'bus.

At Dulesgate on Saturday I found a specimen of *Lepidodendron vasculare* with unequal branching, a very small stele passing off from a normal primary stele. I also got one or two specimens of *Botryopteris hirsuta*.

August 22, 1938. On Saturday last I went to Littleborough and dug up some of the tip at Shore. I brought away a bag full of coal-balls. Amongst them there was *Stauropteris* and *Calamites*.

Today I have been over to Bacup. First I went to see Mr Hargreaves, who offered to take me over to Whitworth (SD*885195*) some Tuesday afternoon. I then went up to Deerplay and afterwards came down to Mr Thomas Ashworth's house where I was asked to tea. He promised to get me some coal-balls from Old Meadows and Nabb. Whilst at Deerplay I saw Mr Hugget the Manager of Towneley Coal and Fireclay Company. He promised to have some coal-balls sent out of Deerplay pit for me.

August 31, 1938. At the weekend I went over to Bacup on Saturday, to Old Meadows and examined some coal-balls which were brought out of the pit a long time since for Duerden but were not sent down to London. They showed the presence of *Lyginopteris, Lagenostoma ovoides, Sphenophyllum, Lepidodendron vasculare*, and *Botryopteris hirsuta* and *ramosa*.

September 8, 1938. Today has been a lovely day. I caught the 2 p.m. 'bus to Walsden Waggon and went up Ramsden Clough. I saw a Small Copper butterfly near Ragby Bridge. The latter had been partially destroyed in the recent storm. Near the old tumble-down Moorcock Inn I saw one or two very tiny birds in some hawthorn trees. I think they were Goldcrests. They were yellow underneath and about 2½ inches long. Two Kestrels came overhead sailing beautifully in the sunlight and against the S.E. wind. I went up Foul Clough and examined the Lower Mountain Mine exposure, satisfying myself as to its correct identity. I looked for but could not find the Upper Foot seam. I then struck across to Knowsley Clough (SD*900204*) where a local man is working the Upper Foot coal privately. He goes in about once a week and has only worked it 2 to 3 years. There are a few "brass lumps" in the coal but no nodules. Nearby are about 3 tips where I found many roof balls. This was like at Buckley (Clough Bottom, SD*905152*).

The man recommended me to go to the back of Hey Farm

near Wardle and to another place also. He worked in a quarry, had a cottage farm, some land and beasts, a wife and two children and got his own coal. The place is very near to Shawforth (SD 892207). I walked back by way of Foul Clough Colliery Road; it was a beautiful moonlight night, very still.

September 13, 1938. On Saturday last I went over to Bacup in the morning, visiting Old Meadows Colliery; it was a lovely morning. I got very little, however, merely a *Stigmaria* axis with a centripetal ring of primary xylem, probably the *Stigmaria* of *Bothrodendron mundum*.

September 19, 1938. On Saturday, though it was a dull day with some rain, I went with a friend to Littleborough. First we looked on some old tips to the east of the town and railway. They looked to be Upper Mountain Mine workings and yielded nothing. We then walked out to Shore and after sheltering for some time borrowed two picks and a shovel from the farmer and dug the tip surface in places and obtained a few coal-balls.

September 22, 1938. At about 10 o'clock this morning I set off to walk over to Wardle. I walked along the canal bank to Gauxholme (SD 929232), then struck up Naze Road. It was very close and this made me sweat. I then made straight for Ramsden Road and continued up to the top of Foul Clough (SD 907202). I had a look at the east end of Rough Hill (SD 912203) for the Upper Foot seam and found indications but nothing more. There was a tip nearby (near the stream) which was possibly due to the Upper Foot seam having been worked at some time though there was no positive evidence *eg* no roof balls or coal-balls. I walked round the west side of the valley and so down to Wardle (SD 913166) but I found nothing. Crossing through Wardle I examined a tip behind the Hey and found a poor coal-ball. This is less than a quarter of a mile from Shore and yet apparently coal-balls are scarce in the Upper Foot seam here. Arriving at Shore about 5 o'clock I dug a number of coal-balls out of the tip until dusk fell about 8 o'clock. It turned out quite a nice day in the afternoon and evening.

September 23, 1938. I have been examining the coal-balls I have collected this week and find that in the Dulesgate material there is a specimen of *Etapteris scottii*. In the Shore material I found *Stauropteris, Calamites, Calamostachys, Lyginopteris,* and three specimens of *Lagenostoma*, one possibly *L. lomaxii* and two *L. ovoides*.

The latter showed the early stages of prothallus formation as figured by Prankerd (1912).

September 27, 1938. On Saturday last I went on the natural history ramble of the Yorkshire and Manchester Geological Societies. We met at 2.15 p.m. at Copperas House and went up Henshaw (SD93–22-) and as far as Bottomley Clough (SD94–21-). Passing Lower Hollingworth Farm (SD93–21-) there was a sale in progress, the auctioneer's voice, musical and sing-song being heard as he knocked down a cow for £18/15s.

Yesterday (Monday) I walked over Sourhall (SD916247), Flowerscar (SD905247), Sharneyford (SD888246) and Deerplay (SD859268) to Nabb Colliery (SD852261) at Water. I set off about 9.45 a.m. On Flowerscar I searched for flints but found none. Arriving at Deerplay I found that no coal-balls had been got out as was promised. I therefore moved on to Nabb where the manager promised to get me some more. They have a great quantity of coal in stock here and have been working short time. They are erecting baths under the Miners' Welfare Scheme. Calling at Nabb Farm the A.R.P. people were just bringing round gas-masks. Coming home I searched the flanks of the moor behind Deerplay pit and found a few flints, my first. During the day I saw *Pilobolus* for the first time. When I got home at night the A.R.P. men (Mr Rushworth and Mr Crabtree) brought round gas masks and I tried one on.

September 30, 1938. Yesterday was a beautiful day, in keeping with the news of the European situation. Monday, Tuesday, and Wednesday were three very trying days. Everyone thought there would be war over Hitler's intended invasion of Czechoslovakia. Today, Hitler, Chamberlain, Daladier and Mussolini have met and come to an agreement. Yesterday I went on the 1.30 p.m. 'bus to Littleborough and walked out to Shore. I dug a number of coal-balls out of the tip. The afternoon was really glorious, true Autumn weather. I dug more coal-balls than I could carry. I examined the material at home and found *Amyelon radicans*, good specimens; *Calamites, Stauropteris, Lepidostrobus* and one rather poor *Ankyropteris westphaliensis*. Someone had been attempting to open the old mine mouth. I think the mine could be entered.

October 2, 1938. Yesterday I went to Shore, Littleborough to collect coal-balls. After digging and getting as many as I could I had a chat with a man living in the cottages near to the farm.

He informed me of three places where he thought the Upper Foot seam had been worked.

October 7, 1938. Yesterday afternoon I went up Dulesgate walking over by Dobroyd. I dug a good patch of the tip surface and got a number of coal-balls though there was not a great deal in them.

October 8, 1938. This morning I went over to Bacup on the 8.20 a.m. 'bus and called at Old Meadows Colliery. Mr Hunter, however, had not got any coal-balls out. I then walked up to Nabb where the manager had very kindly got three trucks full of coal-balls out. I cracked a good number of them and found *Lepidodendron vasculare, Lepidophloios fuliginosus, Sigillaria* sp. vascular axes, *Lyginopteris oldhamia* (certain large nodules were full of this), *Botryopteris ramosa* petioles, *B. hirsuta, Stigmaria bacupensis* and *S. ficoides.* Close on 4 o'clock I left, since it was starting to rain. It rained very heavily and by the time I got to Deerplay Inn I was very wet. I caught the 'bus there and came right through to Todmorden.

October 15, 1938. Today I was up at 7 o'clock. I just missed catching the 8.15 'bus to Littleborough so caught the 8.55 a.m. I arrived at Shore and dug the tip for a time and then walked up towards the reservoir (SD*921175*) near Moor Gate Farm (now empty). Just near the S.W. corner of this reservoir a man has commenced working the Lower Foot coal. He tells me he is unemployed and has worked in Barnsley and also in the Markham pit before the great accident. I walked round the bend of the hill and over to the new Wardle Reservoir (SD*913176*). Here I crawled down a day-hole to find out what seam was there. It was the Lower Mountain mine. Nearby in the newly cut bank, the Lower Foot is also exposed and further on, across a little stream and up behind a barn are some tips possibly of the Upper Mountain mine. I did not find, however, the day-hole of the old Upper Foot mine though I saw it on the reservoir side of the hill. I dug up numerous coal-balls in the afternoon which turned out quite nice and afterwards had a chat with an old cotter at Higher Shore Cottages. He has worked the Upper Foot and I shall try and induce him to take me in the old workings D.V., if possible next Wednesday.

October 19, 1938. Today I have not been up to Manchester. This morning I stayed in and cleaned coal-balls; the most interesting one was from Dulesgate containing what looks like *Botryopteris cylindrica*. One of the Shore specimens contained a large crushed stele associated with *Lepidophloios* outer cortex. Yesterday I was interviewed by Professor Pugh for geology classes in the PhD. course for which I am registering. This afternoon I have been to Shore. I caught the 1.35 p.m. 'bus and got there soon after 2 o'clock. I called at the cottages at Higher Shore where the one-time collier lives and together we went up past Moor Gate Farm to an old day-hole into the Upper Foot mine. We went in for about 400 yards and came out onto an artificially made escarpment overlooking Wardle new reservoir. These old workings in the Upper Foot seam are remarkably good and not too wet. One tunnel which we did not follow went in the direction of the old pit at Shore, and may possibly connect up. The coal was of a poor quality with much dirt in it from the marine band. We found one coal-ball and left it inside since we did not go back the way we came. I came back with a bag full of coal-balls which lads had collected on the tips.

October 22, 1938. This morning I went on the 8.20 'bus to Bacup; it was a slightly foggy morning and worse up Dulesgate. The leaves left on the trees from the recent gales are turning colour; in Dulesgate one or two trees have been blown down.

After obtaining some hydrochloric acid at Boots I went to Old Meadows Colliery and there cracked some coal-balls which Mr Hunter had got from the north (*ie* Deerplay) direction in the pit. I found *Stauropteris* much to my surprise. I left Old Meadows about 12 o'clock and went round the top way to Deerplay. Going down to Nabb I cracked a number of coal-balls and got *Lagenostoma ovoides* and *Sphenophyllum*. I walked back into the centre of Bacup and caught the 6.15 'bus back to Todmorden.

November 5, 1938. Today I have been over to Old Meadows Colliery, Bacup. I went on the 8.20 a.m. 'bus and was there by 9 o'clock. We went into the mine just before 10 o'clock and came out just before 12 o'clock. We brought a number of coal-balls out in which there was *Lepidodendron vasculare*, *Sigillaria* sp., *Lyginopteris*, *Sphenophyllum plurifoliatum* and *Stigmaria ficoides*.

November 9, 1938. This afternoon I went to Littleborough to see Mr E. Clegg in order to ask if there was a possibility of reopening the old mine at Shore. He declined, however, considering the matter impossible.

November 19, 1938. This morning I went over to Bacup to Old

Meadows Colliery on the 8.20 'bus. Mr Hunter took me into the pit again to the same place as on the first two occasions. There are two places quite close together where there are coal-balls. We went to the first and got a good sample of balls. I examined them but did not find them very good and got next to nothing out of them. They are very hard and quite free from pyrites. He tells me there are no more at the place a little northwards towards Deerplay.

November 21, 1938. On Friday last I sent in my application for the vacant post of Assistant Keeper in the Department of Geology at the British Museum (Natural History).

November 23, 1938. I have almost gone through the Manchester Museum coal-ball collection and have started work on the Pettycur material. The latter looks quite promising and certainly contains a rich flora. Already I have seen *Stauropteris burntislandica, Heterangium grievii, Metaclepsydropsis duplex, Lepidodendron brevifolium* and *Botryopteris antiqua.* This material was collected by Dr Kidston and came into the possession of Professor Lang after Kidston's death. Among the Manchester Museum coal-balls were some from Haugh Hill (SD968974) near Stalybridge collected by G. Wild (numbers 1665, 1669, 1675, 1690, 1714); also some from Moorside, Oldham (numbers 1531, 1621, 1633, 1677). Coal-ball 1583 was labelled Trawden, probably the same mine as Laneshaw Bridge near Colne (see Stopes and Watson 1908).

November 30, 1938. On Sunday evening I went up to Manchester on the 10.19 p.m. train. I got a tram down Market Street and got off at London Road Station. Arriving London at 5.55 a.m. on Monday I waited in the Luncheon Room and Great Hall until about 8 a.m. and then went and booked bed and breakfast at the Cleveland Hotel, Montague Street, Russell Square, after which I caught the tube to Hammersmith, and then a 'bus to Kew, arriving soon after 10 a.m. I found Mr Ballard who showed me over the fern houses. He was very proud to show me germinating spores of *Regnellidium* from Brazil which he said had only been found three times before. In the Kew Gardens I saw Mistletoe growing on *Crataegus* (Hawthorn). Everything was still green. I saw an Angle Shades moth sitting on a tree trunk. The following ferns showing buds on leaves were seen at Kew. *Ceratopteris thalictroides, Diplazium proliferium, Tectaria coadunata* var. *gemmifera, Asplenium caudatum, A. lineatum* var. *inaequale* (which had buds on pinnae at bases of pinnule segments), *Woodwardia radicans* (with bud at tip of frond), *Polystichum aculeatum* var., *Hemionitis arrifolia* and *Dryopteris pedata* with bud at base of lamina, *Asplenium gemmiferum* with bud at base of terminal pinna.

I lunched at Kew and then got a Green-line 'bus for Egham. I enjoyed this run out and arrived at the Royal Holloway College at 1.30 p.m. Miss Blackwell came in about 2 p.m. I had 2 hours with Benson's slides of *Botryopteris antiqua*. Miss Blackwell then gave me tea in her study after showing me round the College. It is a magnificent place with lovely grounds. Just as I was there we had a most beautiful sunset. The view from her window was also lovely. For tea we had quince jam, the quince having been grown in the grounds.

Next day I got the tube at Blackfriars and went out to South Kensington. Here I met W. N. Croft in the Elephant Hall and was introduced to Edwards. I then looked at fossil slides of the Scott Collection (mainly for buds on leaves of *Botryopteris hirsuta*), for 2 hours, and was interviewed at 3.30 p.m. I stayed in the B.M. until about 6 p.m. and later got a train at Euston and finally arrived home at Todmorden about 5.30 a.m. Wednesday.

December 20, 1938. Today is Degree Day. Yesterday I found my first specimen of *Botryopteris antiqua* in the Pettycur material. I find I have to etch with both HCl and HF.

December 21, 1938. The degree ceremony went off all right (I graduated MSc).

December 30, 1938. The work that I have done this term has proved that the rachides of *Botryopteris antiqua* produced buds.

January 13, 1939. Tonight I have been at Hebden Bridge, to Croft Terrace, to give a paper to the naturalists on "Plant Movement" but unfortunately no-one turned up except Mr E. B. Gibson. I had an interesting chat with him. Apparently he is named "Binney" after E. W. Binney of Manchester who was a great friend of his grandfather Samuel Gibson. The latter was a Scotsman hailing from the south side of Loch Lomond. He came and settled in Sowerby Bridge as a whitesmith *ie* a maker of ornamental iron work (railings and gates etc) – a sort of aristocratic blacksmith.

Belemnites gibsoni Brown and *Hieracium gibsoni* Backh. were both named after him. He was ornithologist, botanist, geologist, and worked on Mollusca. He also collected moths and some of his collections are now in Peel Park Museum, Salford, and his herbarium at Belle Vue Museum, Halifax.

January 14, 1939. Talking to Professor Lang this morning. He was giving me advice about my forthcoming interview at the B.M. (D.V.) He told me that Foster Cooper has worked on fossil fishes from the Middle Old Red Sandstone of Caithness where he took a farm and reopened a quarry. One of the fishes from there is *Coccosteus*. Professor Lang showed me a good specimen of *Thursophyton* and also a specimen of *Hostimella* with elongated lateral sporangia.

January 19, 1939. Today I have been in London for an interview at Burlington House before the Civil Service Commission for the B.M. post. I went up to Manchester as usual and got to London Road Station at 9.30 We (W. N. Croft and I) arrived at Euston at 1.5 p.m. and went straight out to Piccadilly by tube changing at King's Cross. We had lunch at Lyons and then went straight for interview.

January 24, 1939. I have received word that Mr Croft has got the B.M. post.

January 29, 1939. Yesterday (Saturday) I was in Manchester all day. I went to hear Professor Lang give his Museum Lecture at 3.30 p.m. on "Some Ancient British Land Plants". He mentioned and showed slides of Robert Dick, Hugh Miller and Charles Peach, the three great pioneers of fossil O.R.S. plants.

February 1, 1939. Today Miss Wigglesworth of the Manchester Museum has got a number of books out on Lancashire botanists for the meeting of the North-western Naturalists in Manchester. The work "A Biographical List of Deceased Lancashire Botanists" by Arthur A. Dallman and Margaret Wood, published at York in 1909, has interesting biographical details of Samuel Gibson (1790–1849), his brother Thomas Gibson of Liverpool, John Nowell (1802–1867) of Todmorden and Abraham Stansfield (1802–1880) of Todmorden.

February 12, 1939. Yesterday I went over to Bacup by the 8.20 a.m. 'bus and at Old Meadows Colliery Mr Hunter had very kindly got me more coal-balls from 30 yards or so beyond the union northwards. I got *Botryopteris hirsuta, Lagenostoma ovoides, Calamostachys* and other usual things. In the afternoon I went up Dulesgate and walked up the tips from Cloughfoot to the old South Graine Colliery. At this top end I found a number of quite good coal-balls. The number of roof balls, however, is simply enormous.

February 16, 1939. Today I have been to Burnley *viz* Bank Hall Colliery. The manager got a tub of coal-balls out and I cracked a large number. I got *Botryopteris cylindrica* (good), *B. hirsuta*, and *B. ramosa, Calamites, Bothrodendron mundum, Lepidodendron vasculare, Lepidophloios fuliginosus, Spencerites* and *Stigmaria bacupensis*. It was worth going.

February 24, 1939. Yesterday Professor Holloway from Dunedin, New Zealand, came into the lab. He is staying with Professor Lang. He is a Professor of Botany and Clergyman. He told me he once went and collected coal-balls on an old tip near Halifax. He has been touring the world and has come here from America where he met Fredda Reed. He said there were no fossil plants older than Jurassic in New Zealand.

February 28, 1939. On Saturday I went up Dulesgate in the afternoon walking over Sourhall. I worked right up as far as South Graine. It was fine but with a cold wind. I had a talk to both Temperley brothers. They say they are working the Yard Mine and Little Mine (probably the Lower Mountain and Lower Foot). I went behind their works where there is an exposure of a seam about 3 feet thick.

On Friday last Dr Holloway of Dunedin demonstrated his slides of *Psilotum* prothalli showing embryos and also the presence of a vascular strand with about 3 tracheids, scalariform or annular, and a distinct endodermis and phloem-like tissue.

March 1, 1939. I have now had buds on *Botryopteris hirsuta* leaves from Dulesgate, Shore (not good), Bacup and Burnley. The Manchester Museum coal-balls show them in Halifax material and in coal-balls from near Stalybridge.

March 3, 1939. Today I said goodbye to Mr Croft. I am sorry he is going; he travels Sunday night.

March 4, 1939. This morning I went over to Bacup on the 8.20 'bus and visited Old Meadows where Mr Hunter had again obtained a number of coal-balls for me. I got a few specimens of *Botryopteris hirsuta* but nothing else of value. I think I can regard it as almost hopeless trying to get anything more and must now concentrate on Burnley.

March 9, 1939. Today I have been to Burnley *viz* Bank Hall Colliery. Mr Clegg had got out a wonderful batch of coal-balls but there was not really a great deal in them. It is surprising how barren most of them are especially when compared with Shore material. I got *Lagenostoma ovoides, Lyginopteris* (a little), *Calamites* (root), *Lepidophloios fuliginosus* (common), *Lepidodendron vasculare, Lepidocarpon lomaxii, Botryopteris hirsuta* and *ramosa, Ankyropteris corrugata, Stigmaria bacupensis* (common) and *Stigmaria ficoides.* I have now examined good samples of coal-balls from Nabb; Old Meadows; Bank Hall; Towneley; Deerplay; Dulesgate; Shore; Lightcliffe near Halifax; as well as a few from Crow Edge; Oldham; and Laneshaw Bridge near Colne.

March 13, 1939. Mr Croft sent me a gift of £3 to help towards the expenses of my interviews in London for the B.M. post. It

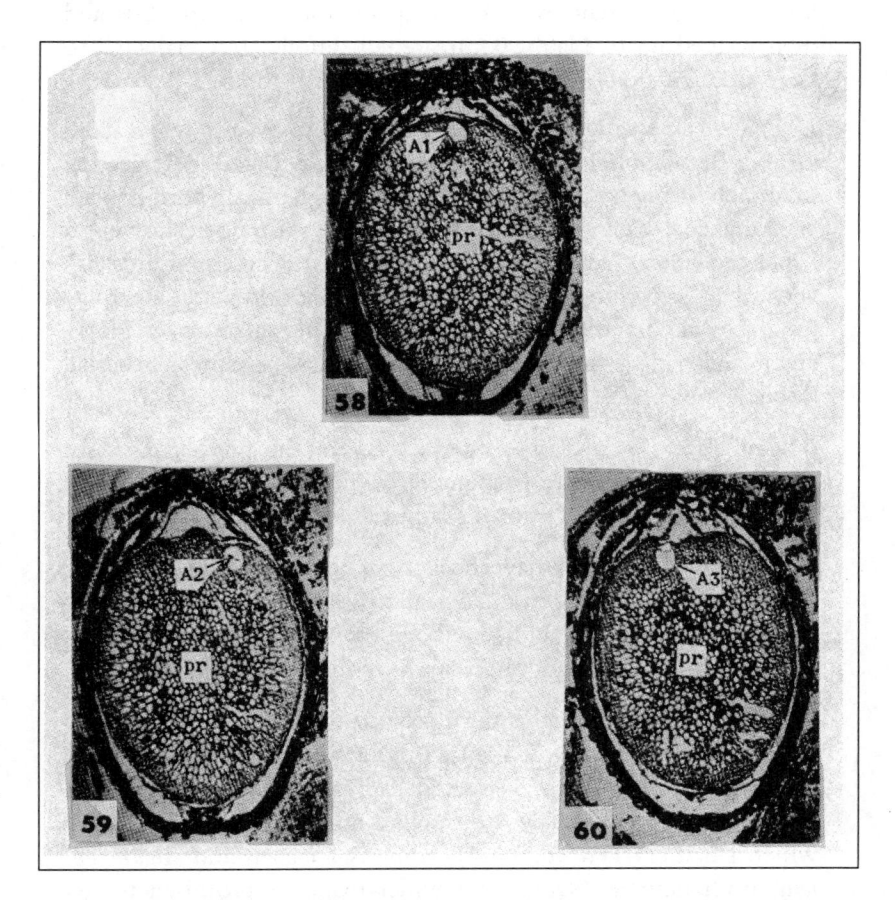

The seed-fern *Lagenostoma ovoides* Will., three longitudinal sections showing a well-preserved female prothallus with three archegonia around the 'tent-pole', from a coal-ball found on Rowley Tip of Bank Hall Colliery, Burnley, Lancashire, and first described in the *Annals of Botany*.

was very kind and thoughtful. He said he had just sold his motor cycle. During the summer he hopes to go to Spitsbergen to collect fossil plants.

April 24, 1939. On Saturday afternoon I went up Dulesgate and collected a few coal-balls. It was very windy. I looked for the Upper Foot seam near Temperley's works but could not find it. *Schistostega* (Shining Cavern Moss) is growing beautifully in mine adits and was in fruit.

May 10, 1939. Today I found another *Lagenostoma ovoides* with a good prothallus. It is not as well preserved as the Dulesgate one.

July 15, 1939. I went over to Bacup on the 8.20 a.m. 'bus and then walked up to Nabb. It threatened thunder but managed to keep fine. At Nabb I was shown over the baths.

July 20, 1939. I caught the 10.45 a.m. 'bus to Bacup and walked up past Broadclough and up the old Deerplay Road. After eating my lunch sitting on a wall I walked over to Nabb and commenced cracking coal-balls. Soon there was a heavy thunder shower so I had to shelter. I did not find much. What a disappointing task it can be. Returning home I branched off the road at Deerplay pit and walked across to Sharneyford. I then walked over Flowerscar and had to shelter during a thunder shower near Sourhall Public House.

July 26, 1939. Today I was interviewed for the post of Assistant Lecturer in Botany at Manchester University. The successful applicant was T. G. Tutin of Cambridge.

July 28, 1939. Today I have been over to Bacup and arrived at Nabb at 12.30 having walked right over. The first coal-ball I cracked was the best. It had a large rachis of *Ankyropteris westphaliensis* in it. I also got *Ankyropteris corrugata* and *Etapteris scottii*.

August 22, 1939. This morning I set out for Nabb shortly after 9 o'clock. On the tops was a thick fog, actually low clouds which wet one's hair through. I walked to Nabb and back again via Sharneyford and Deerplay. I brought a number of coal-balls away which I had cracked previously, some *Stauropteris* and *Anachoropteris*, both new to Nabb. The under-manager promised to get me some more out for next Saturday.

August 26, 1939. I caught the 8.20 a.m. 'bus to Bacup and then walked to Nabb and commenced work on two tubs of coal-balls.

I got *Stauropteris, Lyginopteris, Lagenostoma ovoides, Lepidocarpon, Lepidodendron vasculare, Lepidophloios fuliginosus* and *Calamites*. Soon after 4 o'clock a storm came up from over Waterfoot, with long flashes of lightning. I raced it to Deerplay and there got a lift in a motor-car. It simply deluged; I got it again when back in Todmorden.

August 31, 1939. Again I have been in Manchester and had much discussion with Professor Lang on my *Botryopteris* paper. The political situation is still unsettled. The Navy is being mobilised and school children evacuated from thickly populated areas.

September 3, 1939. On Friday last I went over to Bacup on the 8.50 a.m. 'bus and bought some hydrochloric acid and then walked up to Nabb Colliery. I cracked coal-balls and then had dinner at the nearby farmer's house (Luther King). His brother-in-law, a Baptist minister from Launceston was present.

The political situation has quickly gone from bad to worse, war being declared this morning at 11 o'clock.

September 7, 1939. I am now come to the end of my 5 years at Manchester University. Yesterday I was up for what I intended to be my last day but I still have a few things to bring home. I visited Mr Kerrich M.A. in the Museum and took him some dipterous flies (Tachinidae) which emerged from a larva I got at Inskip.

I had much discussion with Professor Lang and he suggested my going up during the Christmas vacation to complete my paper.

September 18, 1939. On Saturday I walked over to Nabb via Sharneyford and Deerplay. It took about 2½ hours. I collected a few coal-balls and then started back. This was my last expedition for collecting coal-balls whilst a student at Manchester University.

Following is a list of species encountered with brief notes on their occurrence and features of special interest.

ARTICULATALES

1. *Sphenophyllum plurifoliatum*. Localities: Dulesgate; Nabb; and Bacup.
2. *Sphenophyllostachys dawsonii* Will. Localities: Moorside, Oldham; Shore; Littleborough.
3. *Calamites communis* Binney. Localities: most common at

Shore and Oldham, less frequent at Dulesgate; Bacup; and Burnley.
4. *Calamostachys binneyana* Carruthers. Localities: chiefly Shore and Oldham but also at Dulesgate, and Nabb.

LYCOPODIALES

5. *Stigmaria ficoides* Brongniart. Axes and rootlets comprise the bulk of the plant remains in nodules from all localities. There are probably at least two distinct species included under the one name.
6. *Stigmaria lohesti* Leclercq. Localities: Dulesgate; Bacup; and Nabb. Some specimens exceed the size of those described by Leclercq (1925) and Weiss (1928).
7. *Stigmaria bacupensis* Scott and Lang. Localities: Shore; Dulesgate; Bacup; and Burnley.
8. *Lepidodendron vasculare* Binney. Localities: occurs generally at all localities but only sparingly at Oldham and Shore.
9. *Lepidodendron hickii* Watson. Locality: Nabb.
10. *Lepidophloios fuliginosus* Will. Localities: general. Typical axes possess a medullated stele lacking secondary xylem. Smaller axes possess small steles apparently non-medullated.
11. *Sigillaria mamillaris* Brongniart. Locality: Shore.
12. *Sigillaria scutellata* Brongniart. Locality: Burnley.
13. *Lepidocarpon lomaxii* Scott. Localities: Dulesgate; Nabb; Burnley; Shore. The non-integumented form now separated as Achlamydocarpon sp. occurs at Dulesgate; Burnley and Shore.
14. *Lepidostrobus oldhamius* Will. Localities: Shore; Dulesgate; Burnley. Possibly more than one species has been found; those from Shore appear to agree with L. Oldhamius Will. form minor of Arber (1914).
15. *Mazocarpon shorense* Benson. Localities: Dulesgate; Shore.
16. *Miadesmia membranacea* E. C. Bertrand. Locality: Dulesgate.
17. *Bothrodendron mundum* Will. Localities: general.
18. *Spencerites insignis* Scott. Localities: Nabb; Burnley.

FILICALES

(a) *Zygopteroideae*.
19. *Stauropteris oldhamia* Binney. Localities: common at Shore and Oldham; very rare at Bacup, Nabb and Burnley.
20. *Etapteris scottii* P. Bertrand. Localities: Shore; Nabb; Burnley; Dulesgate.

21. *Ankyropteris westphaliensis* P. Bertrand. Localities: Bacup; Nabb; Burnley; Shore.
22. *Ankyropteris corrugata* Will. Localities: Bacup; Nabb; Lightcliffe near Halifax.

(b) *Botryopteroideae*.
23. *Botryopteris cylindrica* Will., now *Psalixochlaena cylindrica* (Will.) Holden. Localities: Dulesgate; Sharneyford; Bacup; Nabb; Lightcliffe.
24. *Botryopteris hirsuta* Will. Localities: general.
25. *Botryopteris ramosa* Will. Localities: general.
26. *Anachoropteris pulchra* Corda. Localities: Shore, Burnley; Nabb.

TERIDOSPERMALES

(a) *Lyginopterideae*.
27. *Lyginopteris oldhamia* Binney. Localities: general.
28. *Lagenostoma ovoides* Will. Localities: general.

(b) *Medulloseae*.
29. *Sutcliffia williamsoni* (Seward). Localities: Shore and Oldham.
30. *Myeloxylon* Locality: Shore.
31. *Trigonocarpus parkinsoni* Brongniart. Localities: Shore: Nabb.

CORDAITALES

32. *Amyelon radicans* Will. Locality: Shore.
33. *Mitrospermum compressum* Arber. Locality: Shore.

Conclusion

As a result of collecting coal-balls over as wide an area as possible in Lancashire and Yorkshire two resultant observations can be stated.

(i) The flora of the Union Mine in the Bacup and Burnley region on the west of the Pennines is identical with that of the Halifax Hard Bed on the eastern side of the Pennines. Both show the relative abundance of *Lepidophloios fuliginosus*, *Lepidodendron vasculare* and *Sigillaria* spp. In addition *Psalixochlaena cylindrica* and *Spencerites insignis* are present but not abundant. The most common fern is *Botryopteris hirsuta* while *Stauropteris oldhamia* is very scarce. Such a flora was probably a climax flora dominated by the arborescent lycopods.

(ii) The flora of the Upper Foot Mine (where separate from

Robin Wood and Whirlaw, from Eagle's Crag, above Todmorden. Photograph by Mr W. Sutcliffe, courtesy of Todmorden Photographic Society.

the Lower Mountain Mine) seen only in Lancashire (*e.g.* at Shore and Oldham) is characterzsed by a higher proportion of herbaceous and shrubby forms such as *Calamites, Sphenophyllum, Stauropteris, Lyginopteris* and Medullosa compared with such arborescent forms as *Cordaites, Lepidophloios* and *Sigillaria* while *Lepidodendron vasculare* is relatively scarce. The impression gained from this Upper Foot flora is that of a transition flora in which there are roughly equal proportions of the arborescent and shrubby (or herbaceous) forms. Coal-balls from the Upper Foot Mine give more well-preserved plants than those of the Union Mine. Thus in one small coal-ball from Shore it is possible to find more and better preserved plants than in a huge coal-ball from the Union Mine. The decay and destruction of tissues seems to have gone on to a much greater degree in the thicker Union Mine. It therefore seems to be axiomatic that on average the quantity and quality of well-preserved plants varies inversely with the thickness of the coal seam in which the coal-balls occur.

REFERENCES

Arber, A., 1914. "An anatomical study of the Palaeozoic cone-genus *Lepidostrobus*". *Trans. Linn. Soc. Lond.* (2) Bot. 8, 205–238.

Hooker, J.D. and Binney, E.W., 1855. "On the structure of certain Limestone Nodules enclosed in seams of Bituminous Coal with a description of some *Trigonocarpons* contained in them". *Phil. Trans. R. Soc. London*, 145, 149–156.

Joy, K.W., Willis, A.J., and Lacey, W.S., 1956. "A rapid cellulose peel technique in palaeobotany". *Ann. Bot.* 20, 635–637.

Leclercq., S., 1925. "Les coal-balls de la couche Bouxharmont des charbonnages de Werister". *Mém. de la Soc. Geol. de Belgique*.

Long, A.G., 1943. "On the Occurrence of Buds on the leaves of *Botryopteris hirsuta* Will". *Ann. Bot.* N.S. 7, 133–146.

. . . . ., 1944. "On the Prothallus of *Lagenostoma ovoides* Will". *Ann. Bot.* N.S. 8, 105–117.

Prankerd, T.L., 1912. "On the Structure of the Palaeozoic Seed *Lagenostoma ovoides* Will". *Journ. Linn. Soc. Bot.*, 40, 461–490.

Raistrick, A., and Simpson, F., 1932–3. "The Microspores of some Northumberland Coals, and the Use in the Correlation of Coal Seams". *Trans. Inst. min. Engrs.* 85, 225–235, and 86, 55.

Raistrick, A., 1934. "The correlation of coal seams by microspore content". *Trans. Inst. min. Engrs.* 88, 142–153.

Stopes, M.C., and Watson, D.M.S., 1908. "On the present distribution and origin of the calcareous concretions in coal-seams known as 'coal-balls'". *Phil. Trans. Roy. Soc. Lond.* 200, 167–218.

Stubblefield, C.J., 1966. "Robert George Spencer Hudson". *Biographical Memoirs of Fellows of the Royal Society*, 12, 321–333.

Walton, J., 1928. "A Method of Preparing Sections of Fossil Plants contained in Coal-balls or in other Types of Petrifaction". *Nature*, 122, 571.

Weiss, F.E., 1928. "On the Occurrence of *Stigmaria Lohesti* Suz. Lecl. in the British Coal Measures". *Mem. and Proc. Manch. Lit. and Phil. Soc.*, 1928–1929, p. 128.

Wright, W.B. and others, 1927. "Geology of the Rossendale Anticline". *Geol. Survey Mem*.

CHAPTER 3

Berwickshire Nature Notes, 1951–59

Reprinted from *The Scotsman* (by kind permission)
with slight emendations and
dates of publication

Hawk Moth Grubs 20.10.1951

The spread of the Rose-bay Willow-herb (or Fire-weed), owing to the cutting down of woodlands in two World Wars, has been accompanied by the spread of a moth whose larva feeds on this plant. The moth is known as the Large Elephant Hawk and is almost as pink in colour as the flowers of its food-plant, though the moth is not to be seen at the same time as the flowers. It is the peculiar larva with eyelike markings which is more commonly seen by country people.

The species was apparently rare in the Borders up to World War II. Thus Mr Geo. Bolam only mentions four records published in his list of Lepidoptera in 1925. During the last four years I have received a number of larvae found by school children in various parts of the county. One was found on a Berwick street. Another came from the Reston district. Two occurred in Duns gardens, and this year one has been brought from Cranshaws. The moth is not seen as frequently as the larva, though I have records from Preston and Gordon.

When living in Staffordshire I reared a batch of about thirty specimens, and discovered that a few were parasitized by a large black ichneumon fly about one inch in length.

The larva is most commonly found in September and early October, and is said to feed also on the Great Hairy Willow-herb (Codlins-and-Cream) as well as Rose-bay.

Unusual Visitor 8.12.1951

An unusual visitor to the Duns district of Berwickshire was a Black

Guillemot. The bird was seen sitting upright and shivering on the banks of the Langton Burn on Sunday, 2nd December. It was in an exhausted condition, and the finder took it indoors in the hope that it would recover. It died, however, before the next morning.

The bird was in typical grey, black and white winter plumage, but perhaps its most striking feature was the vivid orange-red lining to its mouth and beak. One can only explain the presence of such a bird in this locality by assuming that it had been blown inland by the recent northerly gales.

The incident brings to mind another which occurred several years ago in a Yorkshire industrial town. A friend of mine was walking along a street on a winter night when suddenly a bird fluttered down out of the darkness and rolled over at the roadside, dead. He readily identified it as a Little Auk, which had been carried inland some sixty miles by the easterly gale then blowing. It had apparently flown beyond the limits of its endurance, and the little heart had ceased to function.

In a Berwickshire Garden 9.2.1952

Of the birds which visit our gardens, particularly during the hard days of winter, the tits are probably the best known and most popular. Certainly they are energetic, versatile and experimental in trying out new methods of finding food. Earlier this winter in the village where I live, the Blue-Tits intruded into more than one bedroom for the purpose of pulling off wallpaper, and I also saw one in the garden pecking at a garment of underclothing hung out to dry on the clothes-line.

The Great Tit or Ox-eye is also guilty of these exploratory intrusions, and I suspect that it was he who raided our milk bottles during the few minutes they were left on the doorstep. Great Tits are also daily visitors to my beehives, where they pick up any dead or crawling bees about the entrance.

A less well-known visitor to our gardens is the Marsh Tit, which somewhat resembles the Coal Tit, having a black head but no white streak on the nape. Indeed, the bird at first sight looks something like our summer visitor the Blackcap, which may occur very rarely at bird-tables in winter.

I have seen a Marsh Tit fly onto the alighting board of a beehive like an Ox-eye. Frequently, like the Coal Tit (when foraging for beech mast), it will descend to the ground in search of food within a few yards of the observer. I have also watched it removing the white snowberry fruits, carrying them on to the branch of a tree to peck out the seeds. It is resident here in Berwickshire both winter and summer, and I have little doubt that it is one of our regular nesting species.

Spring Moths 29.3.1952

The early spring moths are always of special interest in being the forerunners of so many more of their kind. Here in the Borders our commonest species are probably the Pale Brindled Beauty, the Dotted Border and the March Moth.

They occur in woodland areas and rest on tree trunks during the day. By far the easiest way of finding them, however, is to examine lamp standards and lighted windows after dark. In this way I encountered all three species on the evening of March 18, and a few weeks earlier I saw 13 specimens of the Pale Brindled Beauty on lamp standards in the vicinity of Duns.

These moths illustrate the peculiar vagaries of Nature in the evolution of new characters. Thus here in Berwickshire all the specimens of the Pale Brindled Beauty seem to be pale and mottled, so resembling the lichens which grow on the tree trunks. In industrial areas, however, a dark melanic variety has evolved, and this harmonises better with its sooty man-made environment. Such protective coloration would seem to be of obvious advantage in escaping enemies.

Yaffle Trio 26.4.1952

During the first week of April a young friend took me to see Green Woodpeckers which have recently established themselves in a few places in Berwickshire. The wood we visited (Bonkyl) had been cut down during the late war and is now a desolate heath. Here grow in wild profusion Birches and Willows, Ling and Bell-Heather, Bracken and Gorse, Sphagnum and Polytrichum, and still standing are many gaunt rotten stumps of former trees harbouring the fat white beetle grubs so eagerly sought by Woodpeckers for food.

We were not long among the birch and heather before we heard the laughing call which has earned the name of Yaffle for the Green Woodpecker, but after about an hour's scouting we still had not sighted the owner of the voice. Indeed it was only when retracing our steps homeward through a strip of old Scot's Pines that we got our reward.

Three birds swung out over a ploughed field and then came right into the wood. Their yellow-green plumage stood out against the dark green foliage, and one bird was climbing a tree trunk. It was the first time I had seen the species in Scotland, and I would judge that it is probably here to stay.

Moths and Catkins 24.5.1952

One of the best localities for sallows near Duns is at Kyles Hill on the edge of the Greenlaw Moor. Here is a heathy wood of mixed Birch, Willow, Pine and bushy Heather, a haunt of Roe Deer and many birds including the Tree Pipit.

After dark it was quite a feat to pick one's way by lamplight between the tussocks of last year's Lady Ferns to the "saughs" laden with their golden (male) or green (female) catkins. On these catkins by careful search I found numbers of moths sitting with extended tongues and gleaming eyes as they sipped the nectar.

Some which I found like the Chestnut, Satellite, and the Autumn Green Carpet had hibernated and must have been ready for the feast provided. The Pine Beauty, Common Quaker, Small Quaker, Hebrew Character, Clouded Drab and Red Chestnut had emerged recently after passing the winter as pupae. Swarming amongst the birch bushes was the Mottled Grey occasionally flying to the lantern.

In late April I visited the nearby moor on a hot afternoon for Emperor Moths. Several males were sighted flying at their usual great rate over the heather. As a boy I used to net them, but on this occasion, after much exertion, I discovered that I now required more than a pair of legs for my youthful sport. I therefore withdrew and consoled myself with the hope of visiting the moor in August to catch the bright green caterpillars from which one can rear the moths in captivity, with comparatively much less effort and more success.

Near the Lammermuirs 7.6.1952

During the fine spell of weather in the latter half of May, I visited a quiet little glen at Lees Cleugh tucked away at the foot of the Lammermuirs. Here, by the burnside, are delightful little glades where the blue spikes of the Bugle harmonise with the soft pink of the Rose Campion against a background of uncurling fronds of Male Ferns and Bracken.

Amongst the bracken, we disturbed numbers of the Brown Silver-line moth, and under the Alders we found the delicate Common White Wave, the Bedstraw Carpet and May Highflyer.

Our main object, however, was to see the Pied Flycatchers nesting in their natural haunts among the Alder, Birches and Rowans. Nor were we disappointed. At least six pairs were seen and four nests found, three in Alders and one in a slender Rowan. The hen Pied Flycatcher sits close and may not readily leave the nest. All, however, were readily accessible and one was within a foot of the ground in a fallen tree.

Tree Pipits, Wood Warblers and Garden Warblers were all in song, and Green-Veined White butterflies regaled themselves on the Lady's Smock well called the Cuckoo Flower.

Rails in the Borders 20.12.1952

Recently a pupil of the Berwickshire High School brought me a dead Water Rail. The bird was found on December 14, in a dying condition near the Toll Bridge at Eyemouth and had either met with an accident or was exhausted. Strangely enough, nearby was the head of a Snipe which had also come to grief.

Although the Water Rail is known to have bred in the Borders, it is probable that this specimen was on passage migration. It was in perfect plumage, with grey-blue breast, striped black and white flanks, brown and black back, short tail with white under-feathers and a rather conspicuous red bill.

The specimen showed well the unusual lateral flattening of the body owing to the narrow sternum. This facilitates the bird's skulking movements through reeds and other marsh plants.

The last Water Rail I heard of in Berwickshire was at Grantshouse. It had been caught by a cat which was seen carrying the live bird across the main Berwick-Edinburgh road.

The Water Rail is a cousin of the Land Rail or Corncrake, which still visits the Borders in spite of its reduced numbers. The last Corncrake I heard was on July 12, 1951 at Choicelee, near Duns.

Record of Grey Wagtails 17.1.1953

Of all the birds which frequent our rivers and burns, none excels the Grey Wagtail for grace of movement and delicacy of colouring.

It was about the first week of March that I first observed these birds along the Langton Burn near Duns. By the 22nd pairing flights were taking place and on April 11 nest building had started.

The site chosen was most unexpected – in a wooden garage, used daily, and about 30 yards from the waterside. The first egg was laid on April 20 – a pale greenish-blue egg and rather small. The sitting bird allowed one to admire her delicate grey and yellow attire at close quarters.

Five eggs were laid in all and they were hatched on May 8 – an incubation period of 13 days. The young were duly ringed at one week and left the nest about a fortnight from hatching.

During this period of fledging a pair of Pied Wagtails came and built their nest exactly one yard from the Grey Wagtail's nest and so one had the unusual spectacle of a Grey Wagtail feeding her young while a Pied Wagtail incubated her own clutch of five eggs nearby.

Subsequently the Grey Wagtails built a second nest and raised another family of five, again using a wooden building for a site. This second brood left the nest on June 26.

Since that date I have observed Grey Wagtails in each month of the year – the last date being December 7. None of these birds, however, has possessed rings showing that our young birds had moved elsewhere for the winter. Needless to say, I hope to watch for their return this year.

Berwickhire Mosses 24.1.1953

The Whiteadder, between Abbey St Bathans and Cockburn Mill, makes a wide curve around the northern flank of Cockburn Law, cutting its way through the hard Silurian rocks and forming many deep forbidding pools. At Elba, where there are old copper workings, the river narrows running between steep, rocky banks, which have an interesting and rich moss flora. This was one of the favourite botanising haunts of Dr Johnston, who founded the Berwickshire Naturalists' Club in 1831.

When I visited the spot in mid-January the day was full of the promise of spring. Many of the mosses were in fine fruit, bearing their innumerable stalked capsules ready for a spring sowing of spores. The commonest species was *Rhacomitrium aciculare*, which abounds on rocks and boulders at the waterside. I noted also with interest the brown encrusting patches of *Andreaea petrophila*, a denizen of the hills.

Below Hoardweel, the Dipper was in song, and a Grey Wagtail passed overhead. A solitary Woodcock rose from the river bank, and a large flock of finches came across, from beech trees on the opposite bank. After settling I put the glasses on them; they were mostly Greenfinches, Chaffinches and Yellow Hammers, with one Brambling, and what a handsome bird a Brambling is in full plumage. The sun shone on the blue-black head and brilliant orange scapulars. It was the third time I had seen this welcome visitor in Berwickshire this winter.

Early Insect Life 28.2.1953

The mild spells of weather in January and February have resulted in the emergence of a number of species of moths about a month in advance of last year's dates. Thus the Pale Brindled Beauty put in its first appearance at street lamps near Duns on January 15, and since then I have seen over forty specimens. Their usual date of appearance is about the middle of February.

The evening of January 30 was very mild with slight drizzle, and this brought out the first specimens of the Early Moth. On February

5 I took the first Dotted Border, and after a week of snowy weather on February 14 a single specimen of a Spring Usher came to light. On February 21 I took the first March Moth – again about a month earlier than the date of its appearance last year.

All these five species of moths have females which are practically wingless and cannot fly – like the better-known female Winter Moths which, in spite of this peculiar disadvantage, are sufficiently successful at egg production to provide a major pest for the gardener and a major source of caterpillar food for insect-eating birds like the Robin during their breeding season. Unless there is a severe check, the evidence would therefore suggest an early spring and, judging by the activity of the honey bees, feeding will probably be necessary earlier than usual.

At a Border Lake 28.3.1953

Earlier in the season when the country was wind-swept I decided to pay a visit to the "Hen Poo" at Duns Castle where I knew there would be shelter from the storm.

The lake was dotted with birds. I counted over a hundred Mallard and amongst them several Wigeon. The drake Wigeon is a handsome bird with a chestnut head and lovely yellow stripe over the crown. The underparts are grey and there is a conspicuous white patch on each flank in front of the black tail. A few rose from the water showing the white patch to the front of the wing. More than once I heard their liquid whistling call which one would more readily associate with waders than with ducks.

At the lower end of the lake where there is a sluice-gate a pair of swans appeared to have staked their territorial claims and a dead cygnet was lying at the waterside. Whilst examining the latter I noticed a little white figure on the opposite bank. It was a stoat conspicuous in its winter coat and hunting along the water's edge.

My attention was now drawn to a mixed company of birds which had found shelter and food in the alders and willows nearby. I noted several Blue Tits, Coal Tits and Great Tits and then heard distinctly the "chick-a-bee-bee" of a Marsh Tit. After searching some time with the glasses I spotted it and then to my surprise picked up another bird in smart yellow attire leaning over the branches to peck at the alder cones almost with the agility of the tits. It was a Siskin and later flew away with rapid flight like that of a Goldfinch.

Northumbrian Shore Birds 18.4.1953

There is a marked contrast between the rocky coast-line of

Berwickshire and the flat sandy beaches of Northumberland and in consequence an equally interesting contrast in bird-life.

On March 14 at Fenham Flats opposite Holy Island the sun shone brilliantly on a placid mirror-like sea and conditions were ideal. Several Mute Swans and four Whoopers with their distinctive black-tipped yellow beaks, rode the sea sedately, along with dozens of Wigeon and several Mallard and Eiders. The latter were obviously paired, the black-and-white drakes in sharp contrast to the sombre ducks.

Of the waders the Redshanks were the most difficult to approach – rising and wheeling, turning right and left alternately in perfect unison as they skimmed the glittering water. A few common Snipe and Jack Snipe rose from the shore, the Jacks rising like Skylarks – almost from one's feet. Oyster-catchers and Shelducks were also present in fair numbers and a single Pink-footed Goose. A Grey Plover rose in the distance.

Retracing our steps we came on some large flocks of birds feeding in the wake of the receding tide. There were Dunlins and Knots and a company of Bar-tailed Godwits, the latter having slightly upturned beaks and a direct mode of flight.

The last bird we watched as the mist stole in from the North Sea was a Ringed Plover running nimbly over the beach and then stopping to feed like a listening Mavis on a lawn. Truly the birds of our shores have a charm of their own.

Marsh Titmouse Breeding Places in Berwickshire 13.6.1953

The status of the Marsh Tit in Scotland has long been obscure. According to the "Handbook", the Marsh Tit is replaced in Scotland by the closely related Willow Tit though the Marsh Tit is said to have bred as far north as Etal in Northumberland.

Here in Berwickshire I can say from personal observation that Marsh Tits are not uncommon during winter and early spring. Usually they occur singly or in pairs along with other tits. In the breeding season, however, they seem to disappear.

Early this spring, therefore, I began to keep a sharp lookout for birds showing breeding behaviour in the hope of finding a nest. On April 12 I witnessed what was obviously a territorial quarrel between two pairs of Marsh Tits at Gavinton Bridge, near Duns, but I have failed to locate any nests.

Again on May 16 I saw a pair of Marsh Tits engaged in a dispute with Great Tits near Abbey St Bathans, but again no nest was located. On June 6, however, my ambition was realised when I found a nest with six young in a natural hole in an elm about five feet from ground level. The parent birds went boldly in and out whilst I stood photographing them about nine feet away. Since then a young friend has reported a second nest with young about 30 feet up a tree near the river Whiteadder.

As long ago as 1915, before the distinction between Willow Tit and Marsh Tit was well known, Col. Alex. M. Brown recorded the Marsh Tit in Berwickshire. In his presidential address to the Berwickshire Naturalists' Club he mentioned finding a nest near Longformacus and says, "The Marsh Tit is not uncommon, but is not easily seen". He also adds that he shot a specimen for his collection as early as 1889.

Assembling of Geese 24.10.1953

On a dull afternoon in mid-October I walked with friends across the heathery approach to Hule Moss, in Berwickshire, where every year thousands of Pink-footed Geese take up temporary quarters after their migration from Iceland.

As we approached the larger loch about a dozen Greater Blackbacked Gulls decided to leave. There were also numbers of Blackheaded Gulls and a few Herring Gulls, though nothing like the numbers which congregate from far and near to roost later in the day. One or two Pink-feet stragglers also decided to leave but fourteen other geese resting on the water and probably Grey-lags were unusually tame and decided to stay along with a good company of Mallard, Pochard, Tufted Duck, and a few Shoveller and Teal.

The number of geese seen, however, was somewhat disappointing and after four o'clock we decided to make our way back across the moor. After leaving the loch about a quarter of a mile behind us we heard the unmistakable cry of returning geese, and, turning round we saw the first gaggle coming into view, strung out in a long line against the grey western sky.

Soon this gaggle was followed by others and to the north beyond Big Dirrington came others, and then more, and still more, until within ten minutes the lake had become a centre to which all these living, clamorous streams converged, hurtling down like big Lapwings when once over their target.

By a quarter to five about three thousand geese must have come down onto the water. It was a sight worth waiting for and in the grey dusk we returned with an indelible picture – once seen never forgotten – the assembly of the wild geese at their winter roosting quarters.

Birds of a Burn 28.11.1953

On a three-mile stretch of the Langton Burn near Duns six pairs of

Grey Wagtails acquired territories during 1953. The first pair was seen on February 27 at a nesting site used last year. In three other nests built by different pairs in the mild days of March the first egg was laid on the same day – April 4. Several times I saw the change-over during incubation, one male incubating seven periods (about 3 hours in all) between dawn and dusk.

A pair of Dippers had a nest one quarter built on March 8 but the first egg was not laid until April 16. Another Dipper started laying in the middle of March and reared two broods in the same nest.

Sand Martins nested as usual in the low sandy banks at the burnside but they disappeared early, possibly owing to the severe weather of June.

On August 8 two Green Sandpipers visited the burn on passage migration and stayed for a few weeks, while during the latter half of August we were favoured with visits of a Kingfisher.

At the time of writing the mild weather has brought the Dipper into song and even mating chases have been taking place.

Winter Visitors – Brambling Flocks in Berwickshire 23.1.1954

The Brambling is one of our most attractive winter visitors and yet, to the unobservant, this handsome little finch is often passed by unnoticed. Usually it consorts with Chaffinches forming mixed winter flocks.

The best places for finding Bramblings in winter are shelter belts of old beech trees, especially if they border stubble fields. One such strip near Duns in Berwickshire is known as Buchan's Belt, and here I have watched a flock of these attractive birds several times over the last five weeks.

Bramblings can often be heard before being seen. Their most characteristic note, which betrays their presence, is a rather harsh "tsweek", described sometimes as "scape". In flight they often "talk" to one another with a "chuck" like note rapidly repeated.

The flock I have watched appears to have fairly regular feeding places at certain times of the day. Occasionally they work in small groups under the beech trees, but more often I have seen them fluttering down from the trees into the barley stubble. Their habits (and habitats) seem to be known to our two commonest birds of prey – the Kestrel and Sparrow-hawk, which would also seem to have regular beats.

Four times I have seen a Kestrel visit the Bramblings' field, and twice a Sparrow-hawk has put in a fleeting, dashing appearance. In consequence the birds are very wary and rise with a whirr of silvery wings at the least alarm.

At a spot where the main road passes under the beeches, a Brambling was picked up dead – apparently a road casualty. In the hand one could not but admire the rich orange scapulars and mottled mantle. The orange-red breast is more like that of a Robin than a Chaffinch since it stops half way down. The under parts and rump are silvery white. One also noticed the sharp conical bill, the slender legs, and the tail with a pronounced incut V at the end.

Owls in Berwickshire 13.2.1954

Although the recent frost was severe, one would scarcely have thought it sufficiently prolonged to have serious adverse effects on bird life. I was therefore surprised to receive from a pupil a dead Barn Owl which had been found near Lauder on February 3 in an exhausted condition. Death was probably a result of starvation as when opened up the stomach was very small and empty.

This was the fourth Barn Owl I have seen which had apparently succumbed to cold weather here in Berwickshire. In January 1945 I found a dead specimen on the roadside at Cheeklaw near Duns and another was brought to me by a pupil. The weather at the time was very cold. In May 1946 I found another near the Whiteadder at Cumledge. It had been dead for some time and I assumed that it also was a victim of winter.

Tawny Owls, on the other hand, would appear to be more hardy. The only one I have ever found dead was lying on a window ledge inside a disused kirk in which it had inadvertently trapped itself.

Barn Owls occasionally nest late in the season. Thus, last year I was shown a nest with two young and three eggs in a hollow Elm near the Whiteadder on August 3. There were many pellets in the nest hole and these we were able to extricate by means of a long burdock stem bearing burs. Two dead voles were also present.

Of late years the Little Owl has made attempts to settle in Berwickshire, specimens having been shot in 1949 and 1953.

Crossbills in Berwickshire 27.7.1954

During 1953 Crossbills appeared in Berwickshire in the month of June. The first flock I saw was at Lees Cleugh, in the parish of Langton, on June 28. Since then I have seen Crossbills at a number of well-separated localities, all with coniferous trees *eg* Kyles Hill (August 8), Foulden Hag (August 18), Duns Castle (August 18), Gavinton (October 1) and in Langton estate on several occasions – the last record as recently as February 14.

Usually one is apprised of their presence as a result of their loud excited call notes "jip, jip", made repeatedly in flight. The manner of flight is rapid and recalls that of the Greenfinch. Often they fly at a good height over the tree-tops. At other times they seem to like to perch at the tops of high trees such as Spruce, Larch or Scot's Pine.

I have watched them in the topmost branches of a Larch, tearing off the scales of the cones with their crossed bills so peculiarly adapted to this manner of feeding. Occasionally entire cones were pulled off their stalks and dropped to the ground.

The flocks I have seen have varied from six to fourteen but of late the numbers seem to have been smaller than those seen in summer last year. Crossbills are reputed to be early nesters, and it will be interesting to know if they decide to stay and nest.

Nosema in Bees 10.4.1954

In Berwickshire there is reason to believe that nosema is widespread and probably endemic in many apiaries. Its main method of transmission is probably by the beekeeper himself, aided by the drifting of bees at the heather, together with natural swarming.

Recently I received a sharp reminder of what an effective killer this protozoan parasite can prove. A strong double brood-chamber colony with abundance of sealed light honey was opened up and found completely dead. Bees were lying an inch deep all over the floor and others were hanging dead on the faces of the combs which were heavily fouled with "dysentery". A microscopic examination of the stomachs of a sample of bees revealed the tell-tale spores in their thousands.

It is my belief that the persistence of nosema in apiaries is largely a result of using frames from stocks which have died out in winter.

Birds at Aberlady Bay 20.11.1954

Early in the month I visited Aberlady Bay with friends. We were met by a cold north-east wind blowing from the direction of the Fife coast. The tide was well out but many waders were feeding close inshore. As we made to cross the wooden footbridge a Black-tailed Godwit flew off upstream and a few Redshanks rose with sharp wing beats and musical call notes.

We walked out over the mud-flats dotted with thousands of lug-worm casts. Several Greater Black-backed Gulls were seen and a large flock of Knots wheeled and turned above the shallow water. A mixed company of Curlews, Bar-tailed Godwits and Oyster-catchers were wading in the receding tide near a big stranded tree trunk. A large flock of ducks which were probably mostly Scaup made off farther out to sea.

We saw about two dozen Ringed Plovers – delightful little birds to watch both in flight and when feeding on the sand. A number of very active Sanderlings were feeding with them at the edge of the shallows. Farther out was a small group of Shags, a single Cormorant and a few Shelducks. We walked in the direction of Gullane to a little rocky headland where a few Turnstones were resting on the rocks. Riding the choppy waters here were many Common Scoters, a few Velvet Scoters and a number of Eiders hugging the water's edge.

On retracing our steps we came on a single Snow Bunting which seemed very tired. It scuttled along the line of litter and fluttered close to the ground when approached. Among the weeds above high tide mark were several Linnets, Larks, Meadow Pipits, Starlings and two Goldfinches.

Crossbills in Berwickshire 4.12.1954

As reported previously in these Notes, Crossbills appeared in Berwickshire in June 1953, being first seen at Lees Cleugh. They stayed with us roughly 12 months. Perhaps by that time they had concluded that our climate was not as genial as that found in their own more northerly latitudes.

Most of the birds I saw appeared to be first year juveniles, and it would seem that breeding may not commence until the birds are more than one year old. Nevertheless, I observed the initial stages of breeding behaviour, although there was no nesting to my knowledge.

On April 19 this year I saw a male and female consorting as if paired. The male sang at the top of a spruce "see, see, see, whit, whit, whit, whirr . . ." and occasionally he made gliding display flights from one tree to another, his wings half opened and vibrating.

This cock had vivid red breast and rump, whereas most of the others were less colourful. On another occasion I heard two males singing at the same time and witnessed a short fight.

Altogether fourteen birds frequented the pines at Lees Cleugh and Kyles Hill. The last date on which I saw this flock was June 14. It was about 9 p.m. and the birds had gone to roost in the tops of the pines when we disturbed them. They circled round and eventually settled down once more. Since that date I have not seen another Crossbill in Berwickshire.

Tree Sparrows 25.12.1954

Several times of late my attention has been drawn to Tree Sparrows which are by no means uncommon in Berwickshire though often overlooked. In the village of Gavinton there is a small colony, but owing to their retiring habits the birds are more often heard than seen.

They are fond of sitting at the tops of old trees, bathing in the sun's rays and answering one another with their musical chirps. These call-notes are quite distinct from those of the House-Sparrow, as are the more staccato flight notes ("teck, teck").

On a recent morning when the sun shone brightly for the first time after a week of dull and stormy weather, the notes of the Tree Sparrows were so musical that I had to stop and examine the birds through binoculars before I could persuade myself of their identity. They were obviously enjoying their sun bath in the tree tops.

Only a week ago I came on a similar small colony in old willows by the Blackadder bridge at Nisbet, and a still more numerous flock mixed with Greenfinches at Ladyflat.

In April last I saw a flock apparently resting on migration near the Northumbrian coast at Smeafield. According to Bolam, Tree Sparrows were scarce in Berwickshire last century, but gradually spread inland from the coast.

Early Moths 12.3.1955

It is always a pleasure to see the first specimens of those harbingers of spring, the Early Moth and Spring Usher. Along with the Pale Brindled Beauty, these moths emerge during the first three months of the year.

First dates vary with the coldness of the season. This year the mild weather at the end of January brought them out on the 29th of that month, but no doubt many will still be in the pupal stage owing to the subsequent cold spell.

As usual, males were seen at mercury vapour street lamps which prove a great attraction. Females cannot be observed so easily. Being wingless and incable of flight they must be searched for on tree trunks. A better way of obtaining them is to rear them.

Last June I obtained larvae of the Pale Brindled Beauty by beating birch branches over a beating tray at Lees Cleugh, Berwickshire. So far these have yielded two of each sex. The Spring Usher I obtained in the pupal stage by digging round oak trees at Gavinton and near the Retreat at Abbey St Bathans about the New Year.

Moths that fly across the North Sea 9.4.1955

Recently a Meteorological Office research balloon and parachute descended into my small garden at Gavinton, Berwickshire, the affixed box of scientific apparatus being caught on nearby electricity cables.

The Retreat, below Abbey St Bathans, Berwickshire, from Cockburn Law.
Here I collected moths at night using mercury vapour lamps.
Photograph by David G. Long.

Its place of release proved to be Leuchars in Fife. Such downwind drift of an airborne object illustrates very well what probably occurs in Nature among insects and even spiders.

On August 24 last year I placed an electric lamp and white sheet on the village green after dark when a rough easterly wind was blowing. To my surprise I later found a specimen of the Great Brocade sitting quietly on the cloth.

The following night I treacled telegraph poles near the village and took another. Now the Great Brocade is a rare moth in the Borders, but it is common in the Highlands; therefore I had my suspicions that these moths were migrants.

Later I read in the "Entomologists' Record" that 45 specimens were taken in Westmorland between August 22 and September 2. This is very strong evidence that this species migrated last season.

Incidentally Bolam records how he once found this moth on mussel beds about a mile out from the Northumbrian shore.

Another moth which is both indigenous and migratory is the Golden Rod Brindle. I took three specimens last autumn at Kyles Hill all probably locally bred. The species also turned up in Essex, Surrey, Middlesex and Hertfordshire – far from its breeding haunts. Most of these were the variety *cinerascens* and are considered to have come from the moorlands of central Germany.

April Moths in Berwickshire 23.4.1955

Up to the end of March the cold weather had retarded the emergence of many spring moths, but the warmer days of April soon brought them out.

On the night of April 3 I took my portable mercury vapour lamps to Kyles Hill in Berwickshire and found that the Yellow Horned Moth was at its peak, and about two dozen fresh specimens of this lovely moth came to the lamps. The night was clear, with moonlight, and there was much audible activity among the Pink-Footed Geese on the nearby moor.

A Wood Pigeon was serenading in the moonlight among the pine branches at 10.30 p.m. Other moths in flight were the Mottled Grey, March Moth and the first Hebrew Character was seen.

On April 7 I took my lamps to the same site again, but the night was very different. My diary told me that the moon was full, but no moon was visible, the sky being overcast with thick cloud. At first the moths came well, and then light rain began to fall, and the flight died down.

Suddenly, at about 9.30 p.m., a large black-and-white moth plunged onto the sheet. It was an Oak Beauty, unrecorded by Bolam for the Eastern Borders, a lovely male specimen with typical pectinate antennae. I had already tried for this moth unsuccessfully on two nights in the oak woods below Abbey St Bathans, and now it turned up in a locality where birch predominates.

Other species taken this spring are the Small Quaker, Red Chestnut and Northern Drab (a rarity). These species, along with the Pine Beauty all visit the sallow catkins after dark at this time of the year.

Ravens and Falcons on Cheviot in Spring 7.5.1955

I spent the last day of April on Cheviot with friends. The day was ideal, a blue sky, scudding white clouds, clear vistas, and a bracing wind.

In the College Valley young gorse was ablaze with bloom along the shingle at the river-side. A Green Woodpecker flew across our path and perched on a high rock on the steep hillside. A Kestrel also came along the same hillside, making heavy headway against the stiff wind and finally pausing to rest at the top of an old thorn.

Ascending Hen Hole, we noted Crowberry in flower and here and there odd plants of Wood Sorrel and Wood Anemone. The latter along with Woodrush species (*sylvatica* and *pilosa*), seemed out of place alongside Clubmoss and Butterwort. The moss *Rhacomitrium lanuginosum* covered many a boulder with a downy coat.

Around the tops of the crags a pair of Ravens kept watchful eyes upon us. The only other birds inhabiting this rugged gorge seemed to be the Dippers on the burn. Near the top of the valley we sat down to rest and eat, and were favoured with a visit of a Merlin.

With swift, erratic flight it swung round the opposite hillside and followed the burn to its head, while an audacious Meadow Pipit dashed out in pursuit. Soon after, a pair of Peregrines appeared in the distance and for a few minutes tormented the Ravens with some of their famous mock stoops.

Eventually reaching the top of Cheviot, we found two masses of frog spawn in little pools among the peat hags. We flushed a pair of Dunlins, and one came down among the rough grass and allowed a close approach. Several times it uttered its pleasing low trill. When it rose its small size and low flight made it resemble a Skylark, but the white sides of the rump distinguished it.

There were a few pairs of Golden Plovers about, their mournful notes in keeping with the wild surroundings. Around lay a vast panorama – Hedgehope, the Simonsides, Eildons, Dirringtons, the sea and many a hill we could not name.

Pied Flycatchers 14.5.1955

There are several localities in Berwickshire where Pied Flycatchers return to nest each year in natural sites, sometimes using the same holes in successive seasons. Recently I revisited one of these localities – a secluded wooded dean with an abundance of hardwoods, Oak, Elm, Alder and Birch. And what magnificent birches – some of the largest I have seen in the county, containing at least eighty cubic feet of timber according to one of the woodcutters.

Here I located six cock Pied Flycatchers in song. Though simple, this song holds a peculiar charm for one who anticipates it with each returning spring. "Tree, tree, tree, once more I come to thee".

As I sat and watched one of these birds I saw it visit and inspect three different prospective nest holes, perching motionless outside as when carrying food to its sitting mate.

During my visit I searched the birch tree trunks for the Early Engrailed. This moth was hitherto unrecorded for the county, and as a result of several hours spent in examining several hundreds of trees I got four specimens, two of each sex.

One female I kept for eggs to rear. They are the exact shade of green as the young birch leaves and laid in a cluster surrounded by brown scales from the tip of the moth's abdomen so resembling opening bud scales.

During this haphazard wandering from tree to tree I trespassed on the territory of a Marsh Tit. With protesting notes it informed me that I was intruding, so I retired, and soon saw it fly and perch outside a hole in a rowan. When it left (without entering) I went forward and shone my pencil torch inside the hole. I was met by explosive threats from the sitting hen. Throwing back her head she then jerked it forward and showed her displeasure at my intrusion, refusing to budge. Marsh Tits are widely spread in Berwickshire, and it would be of special interest to know the limits of their range.

Plants of a Burn 21.5.1955

The Langton Burn in Berwickshire cuts its way through the Old Red Sandstone conglomerate and is joined by a tributary burn which makes an even more picturesque wooded gorge below Langtonlees.

In the days of the New Statistical Account a little over a century ago the rare Yellow Marsh Saxifrage, Melancholy Thistle and Stone Bramble were recorded here by the Rev. John Brown, D.D. Unfortunately these plants seem to have disappeared now, although one can never be sure, and there is still much of interest.

It is still the best locality I know in the county for the Prickly Shield Fern (*Polystichum lobatum*), the fronds of which remain green late in the winter. There is also one beautiful patch of Beech Ferns in boggy ground under the Alders. On the dripping rocks in the steepest portion of the cleugh there is Hart's Tongue Fern in company with Golden Saxifrage, Herb Robert, Woodruff and the common liverworts *Conocephalum, Pellia* and *Plagiochila*.

Nearby a clump of Wood Vetch puts up its lovely pale blue flowers each year, and one year I saw a single flowering spike of the Bird's Nest Orchis. This plant grows more commonly under the beech trees at Gavinton, some two miles lower down the burn. Stag's Horn Moss (*Lycopodium clavatum*) and Milkwort grow on the grassy braes and Marsh Speedwell in the sheep drains.

Recently I made another pleasant discovery which quite took me by surprise. At the time I was not on botany bent but was watching a cock Pied Flycatcher when to steady myself I nearly put my hand on a clump of Goldilocks (*Ranunculus auricomus*).

This flower can readily be overlooked as a Buttercup – which indeed it is – but the small flowers and deeply incut leaves betray that it is a distinct species. It favours damp, shady glens and I know of only one other locality for the plant in this county – on the bank of the Whiteadder near Elba.

Migrant Moths 23.9.1955

Recently (September 16) a pupil of the Berwickshire High School brought me a Convolvulus Hawk Moth which was picked up in the centre of Duns, having possibly been attracted at night to the street lamps. This is one of the largest species of British Lepidoptera, and the specimen found measured four and a half inches across the open wings.

It is well known that this moth migrates as it is a powerful flier and as Bolam says, "A Convolvulus Hawk seen hovering at a flower presents about the beau-ideal of energy in an insect, vanishing at one moment like a phantom, to reappear, perhaps a few yards farther off in equally uncanny fashion". Dr Ford in his recent book on moths states, "Nearly all imagines of this species seen in this country will have flown from the neighbourhood of the Mediterranean, a distance which this powerful insect can easily traverse".

A few days earlier (September 11) another pupil brought me a Humming Bird Hawk Moth which had been captured in Gavinton, near Duns. This is another wanderer from the Continent visiting these islands in most good summers. The autumn moths are often descended from larvae bred in this country from immigrant females. It is doubtful whether they can survive our winters although it is known that they attempt to hibernate.

Another migrant which put in an appearance at treacle near Nisbet this month was the Pearly Underwing, a moth which is likewise sporadic in occurrence. It was known in Berwickshire as long ago as 1877, when Simpson Buglass bred a series from larvae obtained on lettuce and cabbage in Ayton Castle gardens.

I am now hoping that the Death's Head Hawk will put in an appearance as it is usually late autumn when it is found. That would be a fitting conclusion to what has been a wonderful season for the lepidopterist.

Snow Buntings 26.11.1955

On October 28 I visited Pease Bay, on the Berwickshire coast. A strong north-easterly gale was piling up the white horses and blowing the frothy spume over the beach like huge snowflakes. In contrast to my last visit in August, there was not another soul about. The Crows and Gulls and a few Redshanks were in complete possession.

I climbed the steep cliffs at the north-west end to get what is almost a bird's eye view of the bay. Beyond the curved sweep of the bay three headlands thrust out their rocky talons into the sea – Greenheugh Point, Siccar Point and Fast Castle. It was a scene well worth the climb. From Pease Bay I went south to Dowlaw. On my way I walked through fields south of Redheugh, where I came on a flock of ten Snow Buntings resting in the lee of a dry stone dyke. They rose into the wind like a flurry of snowflakes, true harbingers of winter.

At Dowlaw, Herring Gulls were following the plough regardless of the tractorman in his cabin, but at my approach they withdrew. I noted a few windblown Blackbirds, possibly immigrants from Scandinavia. Two winters ago I caught one in Gavinton bearing a ring from the Isle of May. I saw no Bramblings at the coast, but a fortnight ago I counted 61 in a straggling flock flying high up over the Hungry Snout in the Lammermuirs. At first I thought they were Linnets until the unmistakable harsh "scape" betrayed their identity.

At a Border Moss (Threepwood) 7.4.1956

On the last day of March I visited a Border moss in Roxburghshire. This is a large undrained area of bushy heather, and was described by Andrew Kelly, an old Berwickshire naturalist, as a "weird place, full of holes". Among the heather grow Cotton Grass, now in flower, Cross-leaved Heath, Crowberry and many mosses. The heather is surrounded by a fringe of scrubby Birch and Sallow bushes, with a few Junipers.

It was my intention to search the Sallows for borings of the larvae of the Lunar Hornet Clearwing moth. I found many Sallows riddled near ground level with old borings, but, strangely, I could not find any showing frass – the sign of active larvae.

The impression I gained, therefore, was that the moth must have been more numerous some years ago than it is at present. In one superficial boring I saw an ichneumon fly searching for its prey, and on several trees I saw evidence of Woodpeckers. Perhaps these enemies have reduced the Clearwing population.

As I walked round the moss I found larvae of the Northern Eggar and Ruby Tiger sunning themselves after hibernation. Soon after I spied a large bird in the distance. As it turned and came towards me, I crouched low and was able to get a good view.

It was a Short-eared Owl hunting by day, with silent wing beats gliding and turning over the heather at about 20 feet. I could make out the facial discs, but the long wings almost gave it the appearance of a Harrier.

After completing my circuit of the moss, two Curlews rose from a rushfield and became very excited. Soon I saw why – a large dark fox was trotting leisurely down the next field, where some tups were feeding quite unconcerned.

When I climbed the fence to leave the moss I accidentally dislodged a moth, which fell to the ground like a bit of lichen. It was the Yellow Horned – a species not easily found by day.

A cock Reed Bunting, with handsome black head and clerical collar, rose from the rushes, and I left it perched on a sallow as if meditating in solitude.

Although this Border moss is set in the midst of cultivation, there is no doubt that it represents a relict flora, and the depth of the peat bears silent witness to an antiquity which must go right back to post-glacial times.

Colourful Burnmouth Braes Flora 9.6.1956

A visit to Burnmouth is well worth while to see the variety of flowering plants which grow on the steep braes looking out over the North Sea.

The Early Purple Orchis known to the local children as "Cockades" is perhaps the most colourful, in marked contrast to the sombre pale green heads of the Salad Burnet. Growing close to the rocks is the Wild Thyme and the first blooms of the gay Rock Rose, while the Wall Pepper or Stonecrop has still only the leaves to show. This latter plant supports the larvae of a rather local species of moth – the Northern Rustic.

Hairy Violets and Cornsalad both grow near the steep footpath and masses of Sloe and the Common Scurvy Grass. The latter has an interesting relative in Berwickshire *Cochlearia danica* with pale mauve flowers and rosette habit. It grows at Siccar Point, but I failed to find it at Burnmouth. Here and there one sees the withered remains of last year's Carline Thistles and the fresh new growth of Rest Harrow and Parsley.

White Wagtails, doubtless on passage, have recently frequented the shingly beach and Housemartins are back at their nesting sites. A bird which still frequents the braes is the Stonechat though its numbers were greatly reduced in the 1947 winter.

Near high tide level many of the rocks are coloured with bright yellow patches of the lichen *Xanthoria parietina* which is one of the main food plants of the hairy yellow larvae of the Dew Footman moth which occurs in little colonies at a few points between Eyemouth and Berwick.

The much larger "hairy oobits" of the Garden Tiger moth also occur feeding on a variety of low plants on the braes. Other larvae occur on the abundant Primroses and Cowslips. The Slowworm, now distinctly local, still survives on these braes.

Another species of plant growing commonly is the Wood Vetch

though not yet in flower. I could not help contrasting its habitat at the coast with its inland station at Langton Lees Cleugh where it grows on the steep sides of the glen amid Hazels and ferns overlooking the tumbling noisy burn, the characteristic haunt of Grey Wagtails and Dippers.

Moorland Nymph in Berwickshire 21.7.1956

In early July, I visited a moorland moss near Greenlaw in Berwickshire. This moss is as flat as the proverbial pancake, though flanked to the north by high ridges or kaimes considered to be of glacial origin. Owing to bad drainage it holds the water like a sponge, and Sphagnum and Cotton Sedge constitute the dominant plants with a mixture of Cross-Leaved Heath, Cranberry and Sundew. Golden Plover, Curlew and Black Grouse find a habitat here to their liking, and it is sometimes tenanted by Black-headed Gulls.

My purpose was to discover if it was also a breeding ground for that interesting moorland nymph – the Large Heath butterfly (*Coenonympha tullia*) and my surmise proved correct. We had not stepped very far on to this open waste before the first butterfly and then others rose and allowed themselves to be carried away downwind where they would either settle or turn and head obliquely upwind.

The Large Heath extends geographically from Shropshire to Shetland, where, according to E. B. Ford, it is probably the only indigenous species.

During our stay on the Moss we also saw the Small Heath butterfly, but it was not nearly so numerous as its larger relative. In one rushy area near the kaimes we also disturbed several specimens of the Small Argent and Sable, a moth which favours moorland ground where Heath Bedstraw flourishes.

Hawk Moths Numerous 22.9.1956

Of the eight species of Hawk Moths recorded for Berwickshire I have seen five during the course of this season.

The Large Elephant and Small Elephant Hawks were both out in good numbers earlier in the year. Thus between May 30 and July 22 I recorded sixteen specimens of the Large Elephant Hawk, and on June 30 I took eight Small Elephant Hawks at the coast near Coldingham. Both these species fly at dusk and two years ago I saw what was probably the same Large Elephant Hawk at honeysuckle in my garden at Gaviton about 10.20 p.m. on three successive evenings.

Our commonest species is the Poplar Hawk, and being a strong flier it ranges from the coast to the hills. I have found its larvae on

Aspens in upland deans and on small Sallows round moor edges. Last year I found one feeding on Birch. The females fly soon after dusk but males are most active in the hour before dawn. Thus, about 2 a.m. on the night of June 14–15, thirteen males of this species visited my lamps on Gordon Moss.

On 7th September, while I was collecting at the Hirsel, near Coldstream, a fine female Convolvulus Hawk dropped into the grass alongside one of my lamps at about 3.40 a.m. Earlier the previous day a pupil of the Berwickshire High School drew my attention to a Death's Head Hawk on the Duns Town Hall sitting about fifteen feet above ground. The same day another pupil found a second specimen at Mossfield, near Grantshouse.

On the following day I searched the lamp standards in Duns, and outside the Horn Inn there was a fine male Convolvulus Hawk sitting very high up. I secured it with the aid of a ladder from the nearby garage. A third Death's Head Hawk turned up in Gavinton on the 12th – unfortunately it had been run over by a car. Finally, a fourth specimen was found inside the British Legion Hut in Duns on the 14th.

Field Mice 4.5.1957

A recent contributor to these "Notes" remarked on the activities and numbers of Field Voles in the Borders. This was evident last autumn and I wonder how long breeding really ceased during the past mild winter.

At the end of March I lifted a roof off one of my beehives at Kyles Hill in Berwickshire and discovered a nest of a Long-Tailed Field Mouse. A feeder had been left on the hive since autumn with a shallow eke to support the roof. The field mice had constructed their nest of grass and leaves between the crown board and roof.

On previous occasions I have found these nests built in October and already occupied by a pair of these attractive mice. It would seem, therefore, that pairs are already formed in autumn.

As I cautiously opened up the nest I was surprised to find inside seven half-grown babies, but all huddled together and dead. Each measured about two inches, including the tail, and possessed a good fur coat and teeth. Their shrunken, dry condition suggested that they had been dead some weeks and at a guess I would say they had been born in February. Had they lived they would doubtless have left the nest by the time I found it.

On the ground at the side of the hive was a collection of empty beech nuts, every one nibbled open at the broad end. With such an abundance of these nuts from last season and the mild, open winter I can well understand the build up in numbers of the Field Mice and Voles.

Pondering the death of this litter of mice I think the most likely cause would be the death of the female parent – probably by falling a prey to some owl or other predator. I have known mice which have entered a hive be killed, presumably by the bees, but with a mouse-excluder on the hive in question this was not likely to be the cause.

Berwickshire Butterflies 27.7.1957

Early in July I visited Gordon Moss by day to see if the Large Heath butterfly could still be found there. Most of my visits to this low-lying Berwickshire bog have been made at night, but a visit by day has its own rewards.

On this occasion rain was falling when we arrived but soon the sun broke through and the humid heat brought out the butterflies and of course the clegs.

Ringlets and Meadow Browns were freshly out and we found several of the former without the characteristic rings on the undersides of the wings. Several Small Pearl Bordered Fritillaries were seen among the Marsh Thistles, Marsh Orchids, Bell Heather and Yellow Tormentil.

Small Heaths were everywhere but its larger relative was not to be found. A female Common Blue was also found frequenting spots where its food-plant grows – the large form of Bird's Foot Trefoil.

Gordon Moss, Berwickshire, from the east, another rich locality for collecting moths. Photograph by David G. Long.

Perhaps the loveliest gem to gladden our eyes was a single Large Emerald moth sitting motionless in its splendour with wings of delicate green open and blending harmoniously with the living carpet of varied herbage beneath the birches.

This season must have been specially favourable for the Lesser Butterfly Orchis of which we saw about 50, their pale cream blooms standing erect among the other marsh plants. In wet hollows Marsh Cinquefoil was in flower and the Sphagnum moss was rearing its globular capsules some of which had burst and released their spores.

Deep among the sedges and rushes the circular leaves of Marsh Pennywort were spotted and by careful examination its tiny heads of inconspicuous flowers were discovered. How unlike the members of the family Umbelliferae with which it was formerly classified.

Habitats such as Gordon Moss are now few and far between in Berwickshire. Long may it remain undrained and undeveloped where Nature's gems may still be found and enjoyed by those who have eyes to see them.

Fossil Plants 3.8.1957

Fifty-six years ago Dr Robert Kidston and Mr A. Macconochie, of the Scottish Geological Survey, visited the Langton Burn near Duns in Berwickshire and discovered plant fossils at a site 400 yards north of the village of Gavinton.

Several times of late I have tried to rediscover this fossil bed but without success. However, a month or two ago, I came on many loose blocks containing petrified plants of Lower Carboniferous age similar to those found by Dr Kidston but their location was roughly 400 yards east of Gavinton.

These plants include very numerous stems of a species of *Lepidodendron* with a solid stele of primary wood; a few petioles known as *Kalymma tuediana*, considered to represent leaf stalks of the seed fern *Calamopitys*; and most interesting of all – a number of "seeds" known as *Calymmatotheca kidstonii*.

Having met with this success on the Langton Burn, I searched the Whiteadder banks between Preston and Chirnside and found a little petrified material below Cumledge and several compressions at Edrom and below Blanerne.

Recently I visited two other fossil sites on the banks of the Tweed at Lennel Braes and Norham Bridge. It was at Lennel that Henry Witham, of Lartington, obtained some of his fossil trees *Pitus antiqua*, described in his book "The Internal Structure of Fossil Vegetables" published in 1833.

Witham was the pioneer investigator of the internal anatomy of fossil plants and it is of interest that he learned his method of grinding sections from William Nicol, of Edinburgh fame.

It is not easy to picture the Southern Uplands of Scotland over 300 millions of years ago, when these plants lived. Much of the land became covered by the sea after the close of the more arid Devonian period, and the succession of shales, limestone bands and sandstones which make up the Calciferous Sandstone series indicate that conditions were probably estuarine or lagoonal.

Fossils are among the most primitive documents of natural history, and he would be a very dull naturalist who failed to sense the fascination of their dumb eloquence after so long a period of total oblivion.

Henry Witham of Lartington (1779–1844). Author of *The Internal Structure of Fossil Vegetables*. Reproduced from *Makers of British Botany*, edited by F. W. Oliver.

Bird-watching in Berwickshire 3.5.1958

On April 8 I visited Allanbank, in Berwickshire, and walked up the right bank of the Blackadder searching the Lower Carboniferous rocks for fossil plants. While examining loose boulders on a shingle bed my attention was diverted by the unmistakable song of a Chiff-chaff coming from some young Alders on the opposite bank of the river. I stopped to watch the bird for some time as it flitted among the low branches making its characteristic jerky movements and uttering its call note. After such a cold Easter it was pleasant to hear this little traveller which had ventured north again in spite of the cold weather.

The following day I visited the coast at Burnmouth and Lamberton and was pleased to see both male and female Stonechats on the sea braes. These hardy little birds manage to find sufficient food by the seashore all through the winter.

Oystercatchers, Curlews, Redshanks, Turnstones, Rock Pipits and

Eiders were all encountered during the day while a few Cormorants passed by flying northwards up the coast. Wheatears were back in the favourite haunts near Lamberton, but the most unusual visitor seen by one of our party was a Water Rail – resting in the old fire-grate of the disused and roofless salmon fishers' shieling.

Another unusual sight I witnessed recently was the method of feeding of a flock of Starlings at Gavinton on April 14. The time was soon after mid-day and the air was appreciably warmer so that many insects were tempted out, including Small Tortoiseshell butterflies. The starlings were flying high up over the village and were obviously hawking flies. At first I thought I was seeing newly returned House Martins, for the manner of flight was identical – a succession of rapid wing beats followed by a short glide and change of direction, quite unlike the usual direct flight of a starling. Evidently the occasion was one worth celebrating, as there were many birds participating in the aerial feast.

A Summer Day in a Berwickshire Valley 19.7.1958

In its almost circular detour through Abbey St Bathans the "White Water" flows to the north of Cockburn Law through a valley thickly clothed with Oak, Birch and Alder of great interest and beauty to those who have an eye for Nature.

Long ago this was a locality for the Pearl Bordered Fritillary, whose food plant, the Dog Violet, grows commonly on the northern slopes of the hill.

Although I kept a sharp look-out for it, all I encountered was a female Painted Lady – a rather tattered wanderer which, along with several specimens of the Silver Y Moth, must have recently invaded our eastern seaboard from the Continent.

This particular Lady was showing great interest in the young thistles, and I have no doubt her intention was to lay eggs. It will therefore probably be possible to find larvae this year in July and August preparatory to the autumn brood of butterflies.

Male Ferns, Lady Ferns and Broad Bucklers were all at their best in the boggy woods below the Law, but most interesting of all were the Beech and Oak Ferns – growing in close proximity on the northern flank of the hill.

From Edinshall Broch the view was magnificent, and no less attractive were the liquid notes of the Sandpipers on the river, the impatient alarm calls of the breeding Curlews and the song of the Redstart in the birches.

Moths that cross the North Sea 18.10.1958

During early October a correspondent forwarded to me for identification a specimen of the Silver-striped Hawk Moth (*H. celerio*) which had been taken by a friend who caught it after it had entered a bathroom by an open skylight in Galashiels. This moth is a native of Africa and Southern Europe, the larva feeding on vines. It is only a casual migrant in Britain and probably the specimens caught in our eastern counties have crossed the North Sea from Holland and Germany.

A specimen of the Striped Hawk Moth (*C. livornica*) was recently reported in the local Press as having been found in Berwick-upon-Tweed. The larva of this species is said to be a pest of vines in Southern Europe.

In September a specimen of the Convolvulus Hawk Moth (*H. convolvuli*) was found at Bridgend, Duns, but unfortunately it had been run over and crushed; probably it had landed on the road as a result of being dazzled by the street lamps.

A specimen of the Death's Head Hawk Moth (*A. atropos*) was brought to me on August 26 from Burnmouth, on the Berwickshire coast. It had settled near the harbour, and had probably crossed the North Sea and then been attracted by one of the harbour lights.

Large numbers of Silver Y Moths have been flying recently both by day and night. These may be offspring of the immigrants which came in May and June. C. B. Williams in his recent book "Insect Migration" states that there is no evidence of winter survival in the latitude of Southern England. Second brood specimens, which I have found in Duns as late as December are said to be sexually immature.

I saw a Painted Lady butterfly on October 1 near the new Berwickshire High School in Duns and several were recorded in September.

Perhaps the most abundant immigrant moth this season and certainly the greatest pest has been the little Diamond-back Moth which has given trouble on Cruciferous crops in both Northumberland and Berwickshire. It is thought that these little moths can rest temporarily like Caddis-flies on the surface of the water when crossing the North Sea.

Bullfinches destroy harmful caterpillars 16.5.1959

Bullfinches are by no means rare birds in Berwickshire although, on account of their secretive habits, they are more often heard than seen. Usually the low clear call notes or a fleeting glimpse of the white rump are all the evidence the bird gives of its presence.

In winter I have seen small flocks numbering about a score, and a fine sight they make feeding on the heather from whence they fly up into the birch trees on the least alarm.

Recently I was much surprised to see a fine cock Bullfinch at close quarters outside the kitchen window where it was feeding in a gooseberry bush. At first I imagined that it was nipping off buds and then I realised that it was actually picking off caterpillars of the Magpie Moth.

These larvae hibernate through the winter, but as soon as the gooseberry leaves are formed in spring the caterpillars recommence feeding and can soon strip a bush of its foliage.

The interesting feature about this incident is that Magpie Moth larvae are reputed to have a bad taste by which, along with their warning coloration, they are protected from birds.

Later we were favoured with a visit of the hen Bullfinch and on three occasions I saw her take a caterpillar in her beak and then drop it. When the cock bird returned I was able to confirm the fact that he actually ate the caterpillars.

CHAPTER 4

Entomological Observations

(a)   Lepidoptera collecting in Berwickshire. A review of past work.
(b)   The Macro-lepidoptera of Northumberland. A review of past work.
(c)   Some Berwickshire Lepidoptera records 1952–1954.
(d)   Collecting in Berwickshire January – June 1955.
(e)   Collecting in Berwickshire July – December 1955.
(f)   Collecting in Berwickshire January – June 1956.
(g)   Collecting in Berwickshire July – December 1956.
(h)   Copper Underwings in Yorkshire.
(i)   The Vestal in Berwickshire.
(j)   The Return of the Orange Tip.

(a) **Lepidoptera collecting in Berwickshire. A review of past work.** *From the History of the Berwickshire Naturalists' Club, Vol. XXXIV. Part 1, 1956, pp. 31–40.*

Since coming to live in Berwickshire in 1945 I have kept records on various aspects of natural history, both botanical and zoological, within the County. Prior to this the Lepidoptera had early attracted my attention and, as a boy, I collected in and around my home town of Todmorden, on the border between Lancashire and Yorkshire. I was fortunate in having a Headmaster at the Todmorden Grammar School who was himself a collector, and gave me much encouragement. Later, I was aided by Mr Harry Britten of the Manchester Museum. He was a field naturalist of the old school, with a wide knowledge, and retained all the freshness and zeal of youth (although well up in years) so that our disparity in age detracted nothing from a common pursuit. He stimulated me to take an interest in some of the other orders of insects, especially the caddis flies, with which he assisted me in the difficulties of identification and mounting.

On arrival in Berwickshire, I found the Club's *History* a mine of information on all things pertaining to the natural history of the area. I soon sensed the spirit of Dr Johnston in founding the Club, and the perusal of his published letters was as if a curtain was drawn aside, revealing something of the intensity with which he persevered with his primary predilection, the elucidation of the natural history of his native region. In his later years his main object seems to have been to publish a *Natural History of the Eastern Borders*, and to this end he undoubtedly encouraged the more youthful James Hardy, who later carried forward the work in such a fine manner. Although Dr Johnston never completed his projected work, his first volume on *The Botany of the Eastern Borders* still remains as a proof of his zeal and ability.

Since Dr Johnston's time (1797–1855) the Watsonian system of Vice-counties has been introduced and used widely and to good purpose, especially for botanical records. Thus for example, in the *Census Catalogues of the British Bryological Society*, one can discover the known distribution by Vice-counties of any species of liverwort or moss occurring in Britain.

This seems to be the most convenient method yet devised for the geographical classification of natural history records, and having it in mind, I have tried to prepare a County List of the Macro-lepidoptera for Berwickshire (Watsonian Vice-county 81). Long ago a similar list, including 335 species was prepared by William Renton for the sister County of Roxburgh, and was published by him in the *Entomologist* for 1903.

In order to make such a list as complete as possible I have searched out old records, mostly in the Club's *History*, and have endeavoured to discover whether the species recorded are with us today. Some species, like the Clifden Nonpareil, *Catocala fraxini*, recorded by William Shaw at Eyemouth in 1876, are probably very rare immigrants and may not be met with again in a lifetime's collecting. Others like the Orange Tip, *Anthocharis cardamines* and Comma, *Polygonia c-album* are probably extinct, though once indigenous. Others, which were formerly indigenous, may still be breeding in the County unknown to us, simply because of the paucity of observers. I refer to such species as the Marsh Fritillary, *Euphydryas aurinia*, and the Scotch Argus, *Erebia aethiops*. The former was recorded on Coldingham Moor in 1850 and 1894 (vol. III p. 5 and XV, p. 223). The latter was recorded for Gordon Moss in 1880 and for Earlston in 1897 (vol. IX, p. 295 and XVI p. 291). Such old records make our work all the more interesting, but it is also true that new discoveries of even greater interest, await the investigator in these less explored fields of natural history. Whole orders of insects still wait for investigation, so that there is no lack of scope for the amateur naturalist. In short much of the material for a work such as Dr. Johnston visualised still remains to be collected by patient observers in the field.

It is indicative of the width of Dr. Johnston's interests that the first recorded list of Berwickshire Lepidoptera in the Club's *History* came from his pen. In 1832, when he gave his first Presidential Address (Vol. I, p. 8) he listed sixteen species of butterflies occurring in Berwickshire. Besides his interesting remarks about the Orange Tip in the Paxton, Swinton and Coldstream area, he also referred to the Speckled Wood, *Pararge aegeria*, as being more common than the Grayling, *Eumenis semele*, on the banks of the River Eye, below Ayton House, in June and July. This remains one of several old records which should be checked at the present day, and illustrates the need for further observations. Another record of Dr Johnston's, which is an even greater mystery, is that of the Silver-studded Blue, *Plebeia argus*. Later P. J. Selby recorded several taken on the banks of the Whiteadder, near Abbey St. Bathans, at a club meeting on 20th July 1853 (vol. III, p. 137). It is strange that when Bolam completed his *Lepidoptera of Northumberland and the Eastern Borders*, he omitted any mention of this species and of the above records.

It must have been soon after the founding of the Club that P. J. Selby (1788–1867) first experimented with a primitive method of "sugaring" for moths. The story has been told by P. B. M. Allan (recently Editor of the *Entomologists' Record*) in his book, *A Moth Hunter's Gossip*, p. 94. According to this author, Selby was not the inventor of "sugaring", but probably got his idea from a note published by Edward Doubleday in the *Entomologists' Magazine* for 1832. Selby apparently communicated his own method in a letter to a Mr. James Duncan, who described it in 1834. Selby's method, briefly, was to anoint a bee-skep with honey and support it over a forked stake about four feet from the ground. He gave a list of species captured in this manner, and it is of special interest to see that he included *Xanthia gilvago*, the Dusky Lemon Sallow which he also included in the Club's *History* (Vol. I, p. 160). This species recorded so long ago by Selby at Twizel, was considered to be of too doubtful identity by Bolam, who only allowed it to stand in square brackets when he made up his list (Vol. XXVI. p. 184). In recent years this species, now known as *Cirrhia gilvago* the Dusky-lemon Sallow has been found in Berwickshire over a wide area. It was first taken at Edrom by Lieut.-Col. W. M. Logan-Home in 1953, and since then I have records from Coldstream, Burnmouth, Nisbet, Gavinton, Kyles Hill and Gordon Moss. Thus I have little doubt that Selby's old record for Twizel was correct. By 1846 it is evident, from a remark made by Selby, that he still practised "sugaring" for moths, although his method had changed slightly to the more usual one of anointing the boles of trees with honey or sugar (Vol. II, p. 210).

In 1843 James Hardy (1815–1898) compiled a list of insects taken near the Pease Bridge. This was communicated to the Club by P.J. Selby (Vol. II, p. 110). In this list *Orthosia gracilis*, the Powdered Quaker is mentioned. This remained the only County record up to 1952, when it was taken at Gordon Moss by an Edinburgh collector (Mr. E.C. Pelham-Clinton). Since then I have also taken it at Gordon Moss and at the Hirsel. Hardy's record, therefore, had to wait 109 years for confirmation.

The greatest outburst of Lepidoptera collecting in Berwickshire seems to have occurred between 1870 and 1880, when about six different collectors contributed records to the Club's *History*. It would also seem that there were other more obscure collectors, like Mr D. Paterson, a Duns cobbler, who recorded *Apatele leporina*, the Miller, from Duns Law in 1873. Mr David Porter, watch-maker, Duns, told me that he remembered this Duns cobbler, whose home was near the Clouds.

One of the chief collectors of that period was William Shaw (1840–1908), an associate member of the Club and a self-taught naturalist, who was at one time a ploughman at Eyemouth. His first published record, made in 1892, was of Camberwell Beauty butterflies, *Nymphalis antiopa*, seen at Eyemouth from 1873 to 1877, and some he also recorded in the *Scottish Naturalist*. He had the luck to take some rarities such as the Silver Striped Hawk, *Hippotion celerio*; Poplar Kitten larvae, *Cerura hermelina*; White Satin, *Leucoma salicis*; Poplar Grey, *Apatele megacephala*; Clouded Brindle, *Apamea hepatica*; Dotted Rustic, *Rhyacia simulans*; Red Underwing, *Catocala nupta* (taken at treacle near Burnmouth); and the aforementioned Clifden Nonpareil, *Catocala fraxini*. His capture of the latter is worth quoting in his own words:

> About the 9th of September [1876], when sugaring near Netherbyres, I was very much surprised to see one of this rare moth. It was sitting with forewings arched upwards, touching each other at the tip, and the hindwings spread backwards and pressing against the tree, and giving this moth a most peculiar-looking appearance. Both the hind-wings were badly torn, but the front wings were pretty perfect. (Vol. VIII. p. 125).

In later life Shaw moved to the Galashiels area, and between 1897 and 1904 published lists of the Lepidoptera of that area (Vol. XVI, p.

231, XVII, p. 87 and XIX, p. 197). In these lists he recorded the Orange Tip, *A. cardamines* and Hornet Clearwing, *Sesia apiformis*, for Gordon Moss. He also recorded the Scotch Argus, *Erebia aethiops*, for Earlston and said that the Juniper Carpet *Thera juniperata* was common around Galashiels. So far there are only two Berwickshire records for the latter species – both in Lauderdale.

Another collector who must have been intimately acquainted with William Shaw was Simpson Buglass of Ayton. Unfortunately, there seems to be no obituary notice concerning him in the Club's *History* and it would be of interest to learn some details of his life. He published four lists from 1875 to 1880 and as many of his captures were made in Ayton Castle grounds it would appear that he was employed there. I have talked to elderly inhabitants of Burnmouth who remember his butterfly collecting on the sea braes. Among his more noteworthy records were the Brown Tail, *Euproctis chrysorrhoea*; Chamomile Shark, *Cucullia chamomilla*; Portland Moth, *Actebia praecox*; Square Spotted Clay, *Amathes stigmatica*; Scarce Bordered Straw, *Heliothis armigera*. He also tells us that he reared a good series of the Vapourer, *Orgyia antiqua*, from eggs obtained in Ayton Woods. To quote his words: "This moth must be more common than we suppose, judging from the old webs on the trees." (Vol. VII, P. 483.) In spite of this statement we have very few definite records of it in the County and it still seems to be peculiarly elusive. It is possible that the habit of the wingless female in laying all her eggs on the old cocoon renders them more vulnerable to the attacks of predators, such as those flies which parasitize the eggs of Lepidoptera.

John Anderson, forester on Bonkyl Estate, was also interested in Lepidoptera about the same time as Shaw and Buglass. His primary interest, however, seems to have been botanical, and he was probably stimulated by contact with James Hardy, who enlisted his help in compiling a list of mosses. He contributed records of Lepidoptera, in 1873, 1874, 1875 and 1889. In 1868 he was made an Associate Member of the Club, along with William Shaw, and after his death in 1893 his brother Adam Anderson (1850–1932), was made an Associate Member. Between them these two brothers made some very interesting records of Lepidoptera, mostly in the Preston area. They recorded the Comma, *Polygonia c- album* and Large Tortoiseshell, *Nymphalis polychloros*, from that district; the Maiden's Blush, *Cosymbia punctaria* from Marygold Hills; Small Clouded Brindle, *Apamea unanimis*, from Broomhouse; Flounced Chestnut, *Anchoscelis helvola* at Hoardweel; September Thorn, *Deuteronomos erosaria*, at Preston; and the Mallow, *Larentia clavaria* at Broomhouse. They also had the good fortune to capture a single

Bedstraw Hawk, *Celerio galii*, "hovering among some flowers" at Preston (Vol. XII, p. 536).

During this same decade 1870–1880, there were at least two other collectors at work in the west of the County, namely, Robert Renton and Andrew Kelly. Again we have only meagre biographical details of these two naturalists, and it is sad that no one has seen fit to leave us an account of what manner of men they were. Robert Renton lived at Threeburnford, near Channelkirk, and published a list of Lepidoptera for that area in 1877.

Later he must have moved to Fans, near Earlston, and in 1880 he published a list of species taken in the neighbourhood of Gordon Moss (Vol. VIII p. 318 and IX p. 295). At Threeburnford, Renton made the only capture of the Large Elephant Hawk, *Deilephila elpenor*, in Berwickshire, known to Bolam when he compiled his list in 1925. Since Bolam's day this moth has increased remarkably, probably as a result of the spread of Rose-bay Willow-herb, which is one of its food plants. When this increase occurred we do not know, though it would seem to have been between 1934 (the year of Bolam's death) and 1944, when larvae were found at Edrom and Coldingham, as reported by Colonel Logan-Home (Vol. XXX, p. 252). Subsequently, larvae were found at Coldstream by A.M. Porteous (Vol. XXXI, p. 54), and I personally have records for Reston, Duns, Gordon, Cumledge, Preston, Grantshouse, Coldingham, Earlston, Cranshaws, Gavinton, Langton Glen, Paxton, Spottiswoode and the Hirsel. Other species of interest which Renton recorded for Threeburnford were the Clouded Buff, *Diacrisio sannio*; Light Knot Grass, *Apatele menyanthidis*; and Ruddy Highflier, *Hydriomena ruberata*. Our knowledge of the distribution of these moths in Berwickshire is still very incomplete. Another little known species recorded by Renton was the Small Yellow Underwing *Panemeria tenebrata* from both Threeburnford and Gordon Moss. It seems that Renton was also the first collector to take the Northern Drab, *Orthosia advena* at Gordon Moss, and I am pleased to be able to say that the species still survives in this remarkable locality. Other records of interest for Gordon Moss made by Renton were the Silver-washed Fritillary, *Argynnis paphia*; the Orange Tip, *A. cardamines*; Scotch Argus, *E. aethiops*; and the Gold Swift, *Hepialus hecta*. According to Bolam (Vol. XXV p. 524) Renton had in his collection in 1881 two specimens of the Purple Hairstreak, *Thecla quercus*, taken some years previously near the foot of the Eildons. This species of butterfly may possibly occur in Berwickshire and should be looked for about oak woods in August.

Andrew Kelly was another Lauderdale collector contemporary with

Robert Renton. He seems to have been both botanist and entomologist as in an article, "On some of the rarer Lepidoptera" (Vol. IX p. 383) he mentions hunting for ferns in the woods round Abbey St. Bathans. He refers to the occurrence of the Speckled Wood or Wood Argus, and remarks that the Wall Brown, *Pararge megera*, appeared to have become scarce. This diminution in numbers of the Wall Brown, between 1860 and 1870, was not confined to Berwickshire, but took place in Northumberland and County Durham. E.B. Ford has remarked upon it (along with the Speckled Wood) in his book *Butterflies* (p. 140). A similar eclipse apparently took place in south Westmorland about the end of the century, and Dr Neville L. Birkett has written that between 1910 and 1935 there were only two or three records in that area. Then, within a few years, a change occurred for the better, and by 1940 the species was widespread and common all over the area (*Entomologists' Monthly Magazine*, 1954, Vol. XC, p. 294). Recently Colonel Logan-Home has informed me that he saw two specimens of the Wall Brown in a Larch wood east of Edrom House, in August 1955. It is possible therefore that this species may yet again increase in Berwickshire.

Another butterfly species which has recently increased its range, after a long period of eclipse, is the Comma, *Polygonia c-album*, concerning which Kelly has another illuminating remark in the article above referred to, where he wrote, "It is very unpleasant to think that *V. c-album* the butterfly of our youth has left us for good and all. It is more than twenty years since I saw it." As the food plant of this species is the ubiquitous stinging nettle there seems to be no reason why it should not extend its range to include Berwickshire. Kelly's reference to the Brown Argus, *Aricia artaxerxes*, is also worthy of attention. He said: "Hartside (near Channelkirk) is perhaps the most wealthy locality in Berwickshire, for this insect." It would be interesting to know if this is true today. Kelly also knew Threepwood Moss near Lauder, and described it as "a weird looking place, full of treacherous moss-holes". In this place he took the Pale Tussock, *Dasychira pudibunda*, a species which is generally believed to be absent from Scotland. Further evidence that it has occurred in our district comes from Bolam, who said that "Shaw had one taken from Greenlaw Moor by D. Anderson, and Renton got it from near Fans" (Vol. XXV, p. 558). It would be of great interest if these old records could be confirmed at the present time. A similar interesting record is that of the Drinker, *Philudoria potatoria* of which he says it occurs on moors, the larvae being more common than the perfect insects. One locality which Kelly made reference to was Langmuir Moss, and I should be pleased if any reader of these notes could inform me of its location. Several times, Kelly made mention of Mr John Turnbull, of Lauder Gas works. He seems to have been another collector about whom little has been left on record. Apparently Turnbull took more than an economic interest in gas lamps, for Kelly tells us that he (Turnbull) took seven gorgeous specimens of the Brindled Ochre, *Dasypolia templi* in 1879 at light. Kelly must have been an energetic man unafraid of night collecting in lonely places, for he recorded catching the Deep Brown Dart, *Aporophila lutulenta* by "sugaring the juniper bushes on the Longcroft Braes". It would also appear that he collected in the vicinity of Allanton, for he recorded the Figure of Eight, *Episema caeruleocephala* from the Blackadder Woods (Vol. VII. p. 234). I have often wondered if Andrew Kelly, the lepidopterist, was also the schoolmaster at Allanton from whom the late John Ferguson of Duns gained his interest in natural history (Vol. XXVI, p. 88).

After the passing of this older generation of collectors, there seems to have been no active resident lepidopterist in the county, and it fell to Mr George Bolam, Berwick-upon-Tweed, to draw together all known records into his fine work, *The Lepidoptera of Northumberland and the Eastern Borders*, the first part of which was published in 1925 (Vol. XXV, p. 515). All succeeding lepidopterists in the county must be grateful for Bolam's work, which provides such a sure foundation for further work in the area. Moreover his list is no dry-as-dust catalogue of names, dates and places, and I, personally, am thankful that he did not spare the printer's ink, but elected to give us such interesting side-lights and anecdotes from his own long experience. It is my only regret that he did not produce more biographical notes on the old entomologists with whom he was personally acquainted, and whose cabinets he had evidently scanned with a careful eye. I specially like his account of bringing home caterpillars of the Fox Moth, as a boy, and releasing them on the strawberry bed to save the trouble of keeping them confined during hibernation. In a similar vein is his reminiscence regarding the lovely larva of the Red Sword Grass, *Xylena vetusta*, "Creeping over the road near Chillingham more than fifty years ago, the carriage in which my mother and I were driving being stopped in order to pick it up" (Vol XXVI, p. 186). In the introduction to his list, Bolam disclaimed any pretension to completeness. In his own words: "Any claim to exhaustivenes is further barred by reason of the fact there still remain vast areas within the district of what may, entomologically, be called *terrae incognitae*, parks and policies, woods and fells, and many inviting upland glens where net has never waved, nor collector ever wandered." This is still true, and as one peruses his list it becomes very evident that the old collectors achieved remarkable results, mainly within their own home districts, but were probably much restricted by lack of transport.

What surprises one is that so much was done without our modern aids. Thus, we read of Andrew Kelly "Sugaring in the woods around Abbey St. Bathans and in the grand old wood of Aiky near Hoardweel". One wonders how he arranged such night excursions in days before even the bicycle was invented. Since the appearance of Bolam's list, various additional records have been made in the Club's *History*. Colonel Logan-Home has made some interesting records of species seen at Edrom, in particular the Tissue, *Triphosa dubitata* and Scarce Tissue *Calocalpe cervinalis* (Vol. XXXI, p 153). From Coldingham have come records, by W.B.R. Laidlaw of some of our rarer butterflies and one new species of moth, the Butterbur, *Hydraecia petasitis*, which has since turned up along the Langton Burn near Gavinton and at Gordon Moss. There is no doubt, therefore, that our knowledge of the distribution of the species of Macro-lepidoptera in the County is far from complete, and what of the insignificant "micros"? With them there is a lifetime's work awaiting someone. Furthermore Bolam's list carries few or no references to many localities which should repay working, places such as the Hirsel, Redpath Bog, Hume, Mellerstain, Legerwood, Corsbie Moor, Bassendean, Spottiswoode, Monynut, Cranshaws, and the small glens like Lees Cleugh, and the Watch valley, near Longformacus. I feel sure that new species are still awaiting discovery and as evidence of this I hope to give some account of my own collecting experience in a future article.

One interesting theory, concerning the fluctuations in numbers of certain Lepidoptera, has been mentioned to me by Colonel Logan-Home, and may well repay observation by field naturalists. I refer to the effects of close cropping by rabbits. Now that the numbers of rabbits have been reduced so drastically by myxomatosis, this biological factor has been largely reduced. It will therefore, be of interest to watch for any changes in our Lepidoptera, correlated with the change in natural vegetation, resulting from the reduction in rabbit numbers. Species like the Wall Brown and Speckled Wood have grass-feeding larvae, which could possibly be affected in this way, and other species may also be affected similarly by changes to other plants. In conclusion I would also suggest that it may still be possible to rescue from oblivion biographical details of some of our former Berwickshire collectors. I refer especially to Simpson Buglass of Ayton, Andrew Kelly of Lauder and Robert Renton of Fans, all of whom have helped to increase very considerably our knowledge of the Berwickshire Lepidoptera.

(b) **The Macro-Lepidoptera of Northumberland. A review of past work.** From the *History of the Berwickshire Naturalists' Club*. Vol. XIV, pp. 18–32.

The earliest records of Lepidoptera taken in Northumberland (Watsonian Vice-counties 67 and 68) are those of John Wallis (1714–1793). Wallis came of a South Tyne family and claimed to have been born at Whitley Castle near Kirkhaugh, about two miles below Alston – just in Northumberland and very near the boundary with Cumbria. He matriculated at Oxford at the age of 18 (in 1733) and later graduated B.A. and M.A. After a curacy at Portsmouth and a spell of school mastering at Wallsend he became curate at Simonburn from 1748 to 1775. Simonburn was the largest and wildest parish in Northumberland, 33 miles from south to north and 14 miles from east to west.

Wallis published his famous book *The Natural History and Antiquities of Northumberland* in 1769 with the help of 294 subscribers. In this book are some of the earliest records of plants and animals occurring in Northumberland. At that time the scientific naming of plants and animals was in its infancy. The binomial system invented by Linnaeus is considered to have started for plants in 1753 with the publication of the *Species Plantarum* and for generic names in 1754 with the publication of *Genera Plantarum*. For animals the start of scientific naming is taken as 1758 with the publication of *Systema Naturae* (10th edition) Vol. 1 though some names go back to *Fauna Suecica* published by Linnaeus in 1746. Thus when John Wallis was writing, scientific names were not really established and he had to use a short description quoting in footnotes Latin descriptions by Ray, Petiver and Linnaeus 1746.

Wallis listed nine species of Northumbrian butterflies –

Orange Tip *Anthocharis cardamines* "the orange-yellow and white butterfly".

Common Blue *Polyommatus icarus* "the small sky-blue butterfly"

Wall Brown *Lasiommata megera* "gold-yellow and brown butterfly which delights much to rest on dry banks, stones and rocks in July and August".

Small Copper *Lycaena phlaeas* – " the small yellowish red butterfly with black spots flying in the latter end of May or beginning of June".

Red Admiral *Vanessa atalanta* "the stately butterfly called the

Admiral – a visitant of gardens and fields in the harvest months". Wallis knew both the larva which he called the *eruca* and the chrysalis and noted that they were variable in colour.

Tortoiseshell. By the Latin footnotes Wallis identified this as the *Large* Tortoiseshell. This was probably erroneous as the insect seen was described as frequenting "Alpine woods". The late Professor J.W. Heslop Harrison thought this butterfly was most probably the Dark Green Fritillary *Argynnis aglaja* (see *Vasculum* 15, 60–62).

Small Tortoiseshell *Aglais urticae* – This is called the Lesser Tortoiseshell and Wallis wrote that it "out-lives the winter by concealing itself in private recesses."

Comma *Polygonia c – album* – "the tortoiseshell butterfly with laciniated wings – not unfrequent in vale meadows and gardens in August".

Peacock *Inachis io* "the peacock's eye butterfly" which Wallis knew could be found in the winter months in close retreats. Wallis did not include the three species of common white butterflies in his list.

Amongst the moths Wallis listed ten Northumbrian species –

Death's-head Hawk *Acherontia atropos* – this he calls the Bee Tiger.

Magpie *Abraxas grossulariata* – "the white, black and yellow moth." Wallis knew that the larvae hibernate and that there is a moorland race feeding on *Erica* and *Vaccinium*.

Cinnabar *Tyria jacobaeae* – "the small beautiful red and dark brown moth". This Wallis had caught in the vicarage garden at Haltwhistle and he knew that the larvae feed on Ragwort.

Burnished Brass *Diachrysia chrysitis* – "the gold-yellow and brown moth". Wallis had found a specimen under an edging of wild thyme in a border of his garden at Simonburn.

White Ermine *Spilosoma lubricipeda* – "the white moth with black spots". Wallis said it was frequent among willows near houses.

Ghost Swift *Hepialus humuli* – " a large white and yellow moth".

He does not mention that the larva is a root feeder known to anglers as the 'docken grub' and used for bait.

Buff – tip *Phalera bucephala* – "the silvery grey-brown and yellow moth". Wallis had received a specimen from "Mrs Reed of Chipchase". He knew the larvae.

Large Yellow Underwing *Noctua pronuba* – "the brown and golden yellow moth". Not unfrequent in gardens.

Garden Tiger *Arctia caja* – "the brown, white and red moth". Wallis knew the hairy caterpillar and wrote, "It is frequent in gardens and has an extraordinary affection for table salads and kitchen greens."

Puss *Cerura vinula* – "the beautiful white and blue moth sometimes observed by the sides of moist groves, under the shade of willows and other aquatic trees but not common." He met with this moth in July 1761 among some tall herbage on the left-hand within the gate in going to Nunwick-Hall.

In the early years of the nineteenth century the most notable Northumbrian naturalists who collected were the brothers Albany Hancock (1806–1873) and John Hancock (1808–1890) Albany was senior to John by two years but died aged 66 while John died aged 82. In their late teens both Albany and John were keenly interested in insects and collected different groups at places like Tynemouth, Prestwick Carr, Winlaton Mill and Gibside. Many of their records are first records for Northumberland and most are for the years 1826 and 1827. They kept similar notebooks planned on the same lines with six columns and similar entries as for example the records for finding and rearing the larvae of the Humming Bird Hawk Moth at Tynemouth. In John's notebook we read – under *Sphinx stellatarum*, "I found 11 caterpillars at Tynemouth in 1826 feeding on the Yellow Ladies Bedstraw, on the 12th August they changed to the pupa and the perfect insect was produced on the 3rd September . . ." In Albany's notebook the entry under *Sphinx stellatarum* reads – "John found on the banks at Tynemouth eleven caterpillars on the Yellow Ladies Bedstraw . . ." He then goes on to describe the colour changes observed when the larvae were preparing to pupate. "They were green with a streak of yellow along each side and minutely dotted with white. A short time before they changed to the pupa they became a dirty pink, the streaks were almost invisible but the dotting remained perfectly distinct. They then covered themselves with a kind of open network where they lay from 12th August to 3rd September when the perfect insect was produced. The pupa was at first white, afterwards dusky."

The first record of the Large Elephant Hawk moth in Northumberland was made by John Hancock at Tynemouth in 1826. Concerning this species there were only two records for Northumberland between 1826 and 1900 and only two records between 1900 and 1941. The sudden increase during World War II must have commenced about 1940. It is now probably as common and widespread as the Poplar Hawk moth.

John and Albany Hancock made the first record for Northumberland of the Gatekeeper or Hedge Brown *Pyronia tithonus* at Hartley and Blyth on 26 June 1826. Albany Hancock recorded the Orange Tip as common at the sides of lanes 4 June 1826 and 29 April 1827.

John Hancock recorded Orange Tips on the Ponteland road on 3 June 1827.

The earliest records of the Small Blue *Cupida minimus* in Northumberland were made at Tynemouth 10 June 1827 and 13 July 1827 by Albany and John Hancock under the name *alsus*. They recorded larvae of the Red Admiral and Painted Lady at Tynemouth in 1826 and a Peacock on 29 April 1827 also at Tynemouth. Albany knew the Dark Green Fritillary and recorded it on the road to Dinnington – on flowers of thistle 1825. They also recorded the Large Heath *Coenonympha tullia* at Prestwick Carr on 15 July 1827.

It is quite possible that about the same period (1826–7) there were other amateur entomologists active in south Northumberland but I have not located published or unpublished historical facts. Certainly in north Northumberland there was a very active naturalist paying attention to the insect fauna about this time. This was Prideaux John Selby (1788–1867). Selby was educated at Durham Grammar School where he had for his contemporaries Sir Roderick Murchison, geologist, and Dr Graham later master of Christ's College Cambridge, and afterwards Bishop of Chester.

From Durham Selby went to University College, Oxford. In 1810 he married a daughter of Bertram Mitford Esquire of Mitford Castle by whom he had three daughters. About the year 1811 he took up residence at Twizel House near Belford. Between 1821 and 1834 Selby published his great work on ornithology with 228 plates of which all but 28 figures were his own work. In 1829 Selby seconded the resolution moved by Sir John Trevelyan which brought the Natural History Society of Northumbria into being. He became a member of the Berwickshire Naturalists' Club in 1832. In 1837–8 along with Dr George Johnston and Sir William Jardine he became an editor of the

Magazine of Zoology and Botany – later the *Annals of Natural History*. He was a Fellow of the Royal Society of Edinburgh and of the Linnaean Society and Durham University conferred on him the honorary degree of M.A. He is buried in Bamburgh Churchyard. Dr James Hardy left a description of P.J. Selby in his unpublished notebook No. 4. This reads, "Mr Selby is an exceeding pleasant, unassuming, polite man, who appears more willing to listen than to speak. He is of middle size, grey haired, mouth and chin rather prominent, with the mouth inward, dressed in plaid suit closely checked and oldish hat, square coat, as had Dr Johnston."

Selby published his work *The Fauna of Twizel* in 1839 in the *Annals of Natural History* Vol. III and part of it is his list of the Lepidoptera taken on Twizel Estate near Belford. Excluding varieties his list includes 255 species of Macro-lepidoptera and 5 recorded on the coast near Belford and Bamburgh.

Among the butterflies Selby recorded the Speckled Wood *Pararge aegeria* and the Wall Brown *Lasiommata megera* and he remarked that the Dingy Skipper *Erynnis tages* was confined to a single field where its foodplant the Bird's Foot Trefoil was abundant. He also recorded the Small Blue *Cupido minimus* in the appendix to his list.

Among moths he said that the Large Elephant Hawk had occurred once – in striking contrast to its abundance today. He recorded the Bedstraw Hawk *Hyles gallii* and got the Convolvulus Hawk *Agrius convolvuli* at tobacco flowers. He said that there was scarcely a tree of the Goat Willow which was not bored by the larvae of the Lunar Hornet Clearwing *Sphecia bembeciformis*.

His records of Noctuidae were much increased by his method of sugaring by the use of honey smeared on a bee skep which "from its circular form allows the moths when settled upon it to be easily captured by the flappers". By this means he also deduced that about three weeks was the average duration of a species in a season.

In his list he recorded the Small Chocolate Tip *Clostera pigra*, the Small Eggar *Eriogaster lanestris* (now probably extinct in the County), the White Satin *Leucoma salicis*, Dingy Footman *Eilema griseola*, Blossom Underwing *Orthosia miniosa* (included only with a ?), and the Uncertain *Hoplodrina alsines* a species recently recorded in south Northumberland (J.D. Parrack).

Selby was able to distinguish and record both the Dark Chestnut *Conistra ligula* and the Common Chestnut *Conistra vaccinii*. The distribution and abundance of these two species in Northumberland is still incompletely known but may be correlated with their food plants. *C. ligula* associated possibly with Hawthorn is the more common of the two at Ponteland but in other parts *vaccinii* appears more common and usually associated with Oak.

Selby recorded the Large Nutmeg *Apamea anceps* under the name *Hama aliena*. We have still only six other records, mainly coastal, for this species in the County. Other interesting records from Twizel were the Double Lobed *Apamea ophiogramma*, Grey Arches *Polia nebulosa* and Sweet Gale *Acronicta euphorbiae* recorded only with a ? The last mentioned has since been recorded by W.G. Watson at Sidwood on 24 May 1919 and was mentioned by G. Wailes in Stephens *Illustrations* 3, 325 so that Selby's records for Twizel could be correct. Similarly Selby's record of the Dusky Lemon Sallow *Xanthia gilvago* was the first for the County; Bolam was sceptical but as the moth has increased and is now widespread I think Selby was right. This may not be so in another hundred years if Dutch Elm disease decimates the Elm population, as the larvae feed on the fruits.

Selby knew the difference between the Common shark *Cucullia umbratica*, and the Chamomile Shark *C. chamomillae*. The latter has been taken in Jesmond (1874, 1899), Dinnington (1962) and Bedlington (1972). This again substantiates Selby's record.

Among Geometers Selby recorded several scarce species such as the Barred Umber *Plagodis pulveraria*, Birch Mocha *Cyclophora albipunctata* recorded under the name *Ephyra pendularia*, Chalk Carpet *Scotopteryx bipunctaria* (rare in Northumberland but common on the magnesian limestone of Durham), Striped Twin-spot *Coenotephria salicata* (recorded as *Cidaria latentaria*), Beech-green Carpet *Colostygia olivata*, Cloaked Carpet *Euphyia biangulata*, Chestnut Coloured Carpet *Thera cognata* – a Juniper feeder, Large Argent and Sable *Rheumaptera hastata* and Grey Spruce Carpet *Thera variata* (though this may have been the Grey Pine Carpet *T. obeliscata* which he did not record).

Amongst the Waves he recorded the Silky Wave *Idaea dilutaria*, Cream Wave *Scopula floslactata* (under *lactata*), Lesser Cream Wave *Scopula immutata* (under *Ptychopoda immutata*) and Plain Wave *Idaea straminata* (under *Acidalia inornata*). It is strange, however, that he does not seem to have recorded the Common Wave *Cabera exanthemata* or Willow Beauty *Peribatodes rhomboidaria*.

From the coast near Bamburgh Selby recorded the Scotch Brown Argus *Aricia artaxerxes* – this must be near the southern limit of its range; also the Crescent Dart *Agrotis trux* (under *A. lunigera*) still the only record for the County and the Bordered Straw *Heliothis peltigera* probably an immigrant. There are a few doubtful species in his list but as his collection went to Cambridge the specimens are no longer available to check.

Another collector who was active in the early part of the last century was George Wailes (d. 1882). He was one of the promoters and founders of the Natural History Society of Northumbria. He drew up the first Trust Deed and carried out other legal matters for the Society. In 1857 he published a list of the Butterflies of Northumberland and Durham (*Transactions Tyneside Naturalists' Field Club*, 3, 189–234). In this paper he mentioned that most of his collecting was done between 1826 and 1834. He included thirty-six species of Butterflies in his list (by modern naming). Like the brothers Hancock he knew the Gatekeeper or Hedge Brown butterfly and said that it occurred in two Northumberland locations –

1. "in profusion in a single field near Whitley where the road to Hartley crosses the Briardean Burn".
2. "Meldon Park – by the roadside about half way between Morpeth and Longhirst". He also recorded the Large Heath at Prestwick Carr and from other Northumbrian sites.

Regarding migration of the Camberwell Beauty Wailes wrote that William Backhouse informed him that about 1820 he saw vast numbers strewing the sea-shore at Seaton Carew both dead and alive. Wailes was sceptical about the possibility of migration and said it was more reasonable to suppose that they had been blown from the land (i.e. Britain) than that they had crossed a sea at least 300 miles wide. For evidence supporting the immigration theory relative to the invasion of 1976 the reader is referred to the article by J.M. Chalmers-Hunt in *Ent. Rec*, 89, 89–105. Wailes mentioned that the Camberwell Beauty was known locally as the "White Petticoats". It seems probable that George Wailes visited Meldon Park near Morpeth and one wonders if that is how John Finlay got his interest in Lepidoptera. Edward Newman quoted a description by Wailes of a visit to Meldon Park in 1831 (see Newman's *Illustrated Natural History of British Butterflies and Moths* pp. 293–294). On the occasion of this visit he witnessed the morning flight of male Antler Moths.

Perhaps the most knowledgeable local entomologist in south Northumberland last century was Thomas John Bold (1816–1874). He was born at Tanfield Lea, County Durham 26 September 1816 and died at Long Benton 5 May 1874 aged 58. Bold lived for most of his life at Long Benton and worked for Mr. T. Pattinson grocer and seedsman in the Bigg Market. At the age of 51 he was paralysed and lost the use of both legs (1867). Earlier in life he was a close friend of James Hardy and together they formed the original Wallis Society when Mr Hardy was running his private school in Gateshead between 1840 and 1846. Bold concentrated mainly on Coleoptera, Hemiptera and

Hymenoptera. In 1877 he related how the Hancock Museum's collection of insects started. A series of drawers was provided. In thirteen of these the beetles purchased from the Rev. R. Kirkwood with a few from other sources were arranged. Names and spaces were provided in three other drawers for Hemiptera and one drawer for Homoptera. To the Lepidoptera eighteen drawers had been allocated. Earlier in the same article he wrote, "We have now a local Entomological Society with a respectable list of members which under the able Presidency of Mr Maling is actively employed in collecting our native insects, devoting, for the present most of its attention to Lepidoptera, of which some good private collections have been formed." After his death Bold's collections were presented to the Society by his brother Edwin Bold.

Although Bold was not primarily a lepidopterist he published records e.g. he recorded a Bedstraw Hawk moth at Newbiggin-by-the-Sea in 1870 and wrote –

> I had the pleasure of seeing this beautiful insect on the wing at Newbiggin on the evening of the 15th of August. An imago . . . was bred by Mr Hamilton (Secretary of our Entomological Club) from a larva found in the engine-shed of the Newcastle and Carlisle Railway on the 7th of September. It fed on Ladies' Bedstraw, pupated and the moth emerged on the 30th of April following. Mr Hamilton thought that the larva had come in sand brought for the use of the engines.

Another noteworthy lepidopterist of the second half of last century was John G. Wassermann. I have no dates for his life-span but he probably died in 1883. He was elected to the Natural History Society in 1871 and his address was 50 Beverley Terrace, Cullercoats. In the Society's *Transactions*, vol. V, 1873–6 (pp. 282–295) he published a paper entitled "Notes on some Macro-lepidoptera occurring at the coast, near the mouth of the Tyne". His collection was donated to the Hancock Museum in 1883 by Mrs Wassermann, suggesting that he probably died about that year.

His records include some interesting species:

> Large Tortoiseshell *Nymphalis polychloros* on an old sugar patch on the palings of his garden 12th September 1875.

> Large Footman *Lithosia quadra*, two, on Westgate and Town Moor, 1872.

> Brown Tail *Euproctis chrysorrhoea*, near Cullercoats.

Yellow Tail *Euproctis similis* at S. Shields.

Wassermann recorded both the Grey Dagger *Acronicta psi* and Dark Dagger *A. tridens* but there is doubt about the latter as the identification was not apparently confirmed by genitalia differences. He said that only the time of appearance could guide the collector in deciding which was which as the Grey Dagger emerges in June and the Dark Dagger in August. To date we do not seem to have an authentic record of *tridens* in the County though Renton claimed to have reared it from three larvae obtained near Kelso, Roxburghshire (see *Entomologist* 1903).

Other coastal species obtained by sugaring were – Reddish Light Arches *Apamea sublustris*, Clouded Brindle *Apamea epomidion*, Crescent Striped *Apamea oblonga*, White Colon *Sideridis albicolon*, Brown Crescent *Celaena leucostigma*, Pearly Underwing *Peridroma saucia*, Stout Dart *Spaelotis ravida*, Square Spotted Clay *Xestia rhomboidea*, Great Brocade *Eurois occulta*, Cloudy Sword Grass *Xylena exsoleta*, Red Sword Grass *X. vetusta*.

William Maling was another keen amateur who took a leading part among the Newcastle entomologists in the latter part of last century. He probably died in 1893 and published lists of Lepidoptera in the *Transactions* between 1870 and 1876. He was elected a member of the Society in 1870 and served on the Committee. His records were later incorporated into Robson's *Catalogue* (published 1899 and 1902). At one time he lived at St. Mary's Terrace opposite the museum and later at 15 Jesmond Road.

John Finlay was another active amateur lepidopterist last century though again I have no dates for his life-span. He died prior to 1902 but was still alive in 1895. This can be inferred from the following facts. J. E. Robson referred to him as the "late John Finlay" in the introduction to the Catalogue. In the *Transactions of the Tyneside Naturalists' Field Club* Vol. 13, p. 173 we read that the first meeting of the Club in 1895 was on 23rd May and was held at Morpeth, Wansbeck Valley, and Angerton – "Meldon House and gardens were enjoyed under Mr Finlay's guidance who also exhibited his unrivalled private collection of Moths and Butterflies which deserved more attention than we could give." John Finlay was apparently the gardener at Meldon House the home of Mr Clayton Swan. After his death Finlay's collection was gifted to the Hancock Museum by a Mrs. Moffat. Unfortunately, Finlay did not put data labels on his specimens. Finlay took the only specimen of the Copper Underwing yet caught in Northumberland but owing to lack of data labels one cannot be sure of the specimen. He also took a specimen of that widely travelled immigrant the Small Mottled Willow *Spodoptera exigua* (Meldon 5.8.1879) as recorded in his diary, where he also recorded the Oak Lutestring *Cymatophorina diluta* three times between 8 and 23 October 1879. The only other record for this species in Northumberland is in Jesmond in 1899 though Bolam recorded it at Alston within a mile of the Northumbrian boundary (Sept. 1917). Finlay knew Greenleighton Moor as a locality for the Large Heath Butterfly and also collected at Coal Law Wood where he got the Early Grey *Xylocampa areola*. This species was still present in this locality in the late 1960s (H. T. Eales).

John E. Robson lived at Hartlepool and died on February 28, 1907 aged 74 years. The first two parts of his Catalogue covering the Macro-lepidoptera of Northumberland and Durham were published in the *Transactions of the Natural History Society* in 1899 and 1902 (old series Vol. XII). The parts covering the Micro-lepidoptera were completed in 1913 (after his death) by John Gardner. His list records a total of 422 species of Macros for Northumberland. For these he relied mainly on John Finlay, J.G. Wasserman, W. Maling and George Bolam.

John Gardner (1842–1921) was born at Egglestone, Teesdale but lived most of his life in Hartlepool. He was a timber merchant and saw-miller. He was both a coleopterist and lepidopterist and knew John Sang also a micro-lepidopterist. After J.E. Robson's death in 1907 Gardner helped to complete the second part on the Micros published in 1913. Otherwise he did not publish much himself apart from records sent to Barrett, Tutt, Fowler and Buckler. His collections of Coleoptera and Lepidoptera came to the Hancock Museum between 1913 and 1921. He died at the age of 79 (21 July 1921). For an obituary see *Vasculum*, VIII, 27.

In this period spanning the two centuries there lived mostly outwith the County the most lavish patron of the Lepidoptera the country has produced. This was James John Joicey who died in 1932 at the age of 61. He was a scion of one of Northumberland's most notable families and had a boyhood interest in insects. Some of his British specimens in the Hancock Museum such as the Black Rustic *Aporophyla nigra* came from the Ingram Valley and are labelled Linhope where he used to collect when on holiday in the early years of this century. From this locality came a specimen of the Triple Spotted Clay *Xestia ditrapezium* the only Northumbrian specimen in the Museum.

When nearly 40 years of age J. J. Joicey embarked on the formation of a private museum at The Hill, Witley, near Guildford in Surrey. His collection of British Macro-lepidoptera was gifted to the Hancock Museum in 1934 and three special cabinets were purchased to house them. The collection was arranged in the cabinets in 1949 by H. Hargreaves with the aid of a Carnegie grant. Obituaries of J. J. Joicey are published in the *Entomologist* 1932, vol. 65, 142–4 and in *Ent. Record* Vol. 44, 68.

Of Northumbrian lepidopterists linking the Victorian generation with those of the present century the best known was probably George Bolam (1859–1934). Bolam's family were of Norman descent. He was born 8.11.1859 at Barmoor but from 1864 lived at Weetwood Hall. In 1877 the family moved to Berwick and he entered his father's office to learn the business of a land agent. He was well acquainted with the Hancock brothers, James Hardy and Abel Chapman. For two years (1906–8) he lived in Wales and wrote a book "Wild Life in Wales". In 1912 he settled in Alston and became very friendly with J. E. Hull the Vicar at Ninebanks and student of spiders. Along with Hull, Bagnall and Harrison he helped to start the *Vasculum* in 1915. Bolam's greatest contribution to local entomology was his work "The Lepidoptera of Northumberland and the Eastern Borders" published 1926–30 in *H.B.N.C.* vols. 25–27. Bolam collected much at Berwick, Kyloe, Haggerston and Newham Bog. From the latter he recorded the Dark Bordered Beauty *Epione paralellaria* 29 August 1890. He used to get the caterpillars of the Confused *Apamea furva* "at the roots of tufts of grass, growing from the sides of the old Berwick town walls in May and June." The moths he took at sugar "at a considerable elevation upon the hill ridges at Langleeford in August and September" proving that Bolam did not confine his moth hunting to places of easy access.

Bolam recorded the Small Engrailed *Ectropis crepuscularia* at Kyloe wood 17 May 1896 though John Finlay had already recorded it in his diary for Stobtree Whin on 24 May 1879. It is now known from six grid squares in the County but Bolam's record is the only one for VC. 68.

Another keen amateur contemporary with Bolam was John Robert Johnson (1865–1935). He was a manual instructor and machine construction teacher at Gateshead Secondary School. In his spare time he made a special study of the life-histories of our three species of Fritillary butterflies and of the small Geometers known as Pugs whose larvae feed inside flowers. He succeeded in rearing the Valerian Pug *Eupithecia valerianata* from flowers of Large Valerian growing by ditches near Prestwick Carr. He was honorary curator at the Hancock Museum between the Wars and for some time put on a monthly exhibit of local butterflies and moths. He was well acquainted with Professor J.W.H. Harrison and Robert Craigs of Catcleugh. For a portrait see the *Vasculum* vol. XXI (1935) opposite p. 119.

During the early part of the present century interest in entomology gradually increased among the Newcastle naturalists. Among those involved was Frederick Charles Garrett D.Sc. a chemistry graduate of Manchester University who became a lecturer in chemistry at King's College, Newcastle. He lived at Hexham and later at Alnmouth where he died 19 December 1940. He took an active part in founding an entomological section of the Natural History Society in 1920. This became the second Wallis Club in 1923. In 1924 he helped to found the Northern Naturalists' Union. Dr Garrett collaborated with Professor J. W. H. Harrison in experiments on industrial melanism in the Early Thorn *Selenia dentaria* and Engrailed *Ectropis bistortata*. It was claimed that melanism in these species had been induced by feeding the larvae with leaves contaminated with manganese and lead salts. This has never been conclusively verified so that it is probable that the melanism was caused by recessive genes in the original stock. Dr Garrett's collection came to the Hancock Museum in 1951 and contains many fine bred series of Lepidoptera from Durham, Northumberland and Cumberland. Melanic specimens of the Early Thorn bred by Garrett are in the Nicholson collection.

In the period between the Wars there lived in Redesdale an amateur lepidopterist known as Robert Craigs. His death probably occurred in 1948. He lived at Reservoir Cottages, Catcleugh and worked for the Newcastle and Gateshead Water Company. His collection of local Lepidoptera came to the Hancock Museum in 1957. Initially he was interested mainly in the bird-life at Catcleugh in the early nineteen twenties and thirties. About 1922 he became interested in Lepidoptera partly through a visit to Rev. J. E. Hull then at Belford. This he described in an article entitled "Lepidoptera in Upper Redesdale" (*H.B.N.C.* 30, 147). In this article he recounted various incidents and mentioned the capture of certain species including the Large Heath *Coenonympha tullia*, Great Brocade *Eurois occulta*, Treble Lines *Charanyca trigrammica*, Golden Plusia *Polychrisia moneta*, Smoky Wave *Scopula temata*, Manchester Treble Bar *Carsia sororiata*, Tissue *Triphosa dubitata*, Thyme Pug *Eupithecia distinctaria* and the Cloaked Pug *E. abietaria*. He also claimed that the Red Sword Grass *X vetusta* is more common about Catcleugh than the Cloudy Sword Grass *X. exsoleta*. Robert Craigs knew the Newcastle collectors such as J. R. Johnson and G. T. Nicholson.

George T. Nicholson lived in Fenham, Newcastle and died 10 February 1949. His collection was gifted to the Hancock Museum in 1951. He was a member of the entomological section of the Natural History Society and later of the second Wallis Club. He was acquainted with J. R. Johnson, Harry Sticks, the Rosies and Professor J. W. Heslop Harrison. He was the first collector to take the Scarce Prominent *Odontosia carmelita* in Northumberland (in the Corbridge area). He reared a lot of specimens including Pugs and frequently collected at Prestwick Carr. He also collected with Robert Craigs in Redesdale. His notebook came to the Museum in 1973. He took the only specimen of the Beautiful Brocade *Lacanobia contigua* yet recorded in Northumberland on 23 June 1910. It was sitting on the trunk of an Alder by the Coldgate Burn in Langleeford Valley.

Professor John William Heslop Harrison (1881–1967) lived at Birtley, County Durham and died 23 January 1967 aged 86. He graduated at Armstrong College in 1903 and became a science master at Middlesbrough High School up to 1917. Later he became a lecturer in Zoology at Armstrong College (1920), Reader in Genetics (1926) and Professor in Botany (in 1927) until his retirement in 1946. As well as studying the Lepidoptera he took an interest in other insect orders and in Botany and was a general field naturalist. He carried out experiments on induced food preferences in sawfly larvae. He added many new species records to the faunal lists of Durham and Northumberland as well as the Hebrides. He recorded the Dark Bordered Beauty *Epione paralellaria* along the Fallowlees Burn in July 1952 (*Vasculum* 37, 24). One of his most valuable contributions is an account of the differences between the various *Oporinia* species (now *Epirrita*) in the *Transactions of the Northern Naturalists' Union* – showing how to distinguish the eggs, larvae, pupae and imagines of the November moth *E. dilutata*, Autumnal moth *E. autumnata*, Small Autumnal *E. filigrammaria* and Christy's Carpet *E. christyi*. This work stimulated more interest in these moths which were further studied by Mr J. Percy Robson, schoolmaster of Barnard Castle. Robson's researches are published in a paper entitled "Variation in the November Moth" (*Ent. Gaz.* 7, 199–200). He was one of the earliest northern lepidopterists to record the Marsh Square Spot *Diarsia florida*. This he had unwittingly taken in Westmorland in 1914 (*Ent. Gaz.* 3. 43). Numerous other notes were inserted in the *Vasculum* mostly relating to Yorkshire and Durham Lepidoptera. J.P. Robson died in 1958 and his magnificent collection was gifted to the Hancock Museum in 1959.

In the period following World War II a prime impetus to Lepidoptera collecting was given by the development of the Robinson mercury vapour light trap. This was later adapted at Rothamstead and used widely especially by participants in connection with the mapping scheme of the Monk's Wood Biological Records Scheme. One of the first to adopt this new collecting technique in Northumberland was

F. W. Gardner B.A., A.M.I.C.E. He was an Army Major who worked for Parsons and had a long connection with the Natural History Society of Northumbria to which he was elected in 1919 and later served on the Council. When in Newcastle he lived in Heaton but after retiral went to Riding Mill where he continued to collect Lepidoptera. Eventually, he moved to Brockenhurst in the New Forest always an attractive area to Lepidopterists. While he was at Riding Mill I corresponded with him as a result of reading his published list "Macro-lepidoptera in Northumberland" (1962) in the *Entomologists' Gazette* Vol. 13, 22–30. Amongst other things he recorded for Riding Mill was a Brimstone butterfly *Gonepteryx rhamni* "flying across the garden in June 1950". This species is resident in Cumbria where its food plant grows but is only known as a stray vagrant in Northumberland. He also recorded the Poplar Kitten *Harpyia bifida* as occasional, the Four Dotted Footman *Cybosia mesomella*, Scarce Prominent *Odontosia carmelita*, Grey Arches *Polia nebulosa*, Double Lobed *Apamea ophiogramma*, Sprawler *Brachionycha sphinx* of regular occurrence, Burnet Companion *Euclidia glyphica* infrequent, Scallop Shell *Rheumaptera undulata* not uncommon near Corbridge, both species of the Lead Belle *Scotopteryx mucronata* ssp. *umbrifera* and *S. luridata* ssp. *plumbaria* (more common), the Common Bordered Beauty *Epione repandaria* and Satin Beauty *Deileptenia ribeata*.

During the period before and after World War II Robert H. Benson lived and collected in Newcastle at Jesmond Park East. He joined the Natural History Society in 1934 and died c. 1957. He was a member of the Amateur Entomologists' Society and his collection was given to the Hancock Museum in 1957. It consisted of life-history stages-ova, larvae, pupae and imagines of species collected locally with a few from the Lake district and Kent where he was apparently stationed during the War. His earliest records were in 1933 and the latest in 1956.

One name fairly well known among amateur entomologists was that of Rosie. The family included at least four different collectors.

1. David Rosie who became a member of the Natural History Society in 1897.

2. Miss Annie Rosie who gave a variety of the Small Copper butterfly to the Museum in 1896. Their collection of Lepidoptera was gifted jointly in 1936. David Rosie successfully practised the technique of preserving and mounting larvae of the Lepidoptera during the period 18898–1902.

3. Alexander Rosie gifted to the Museum a collection of Micro-lepidoptera in 1915 and in 1921 he arranged John Gardner's collection of micros in a new cabinet.

4. Donald Rosie gave two small collections of Diptera to the Museum in 1915 and 1916. In 1950 he presented some Macro-lepidoptera taken in his garden in Newcastle.

Since the passing of former generations of lepidopterists records have been continued and new species added to the County list. The Hancock Museum has thus been enabled to build up a card index of all known records ultimately to help the production of distribution maps.

From the Cheviot area, records have come of the White Underwing *Anarta melanopa* (W. M. Logan Home, I. & B. Wallace *Ent. Rec.* 87, 159), Northern Dart *Xestia alpicola* (M. R. Young *Ent. Gaz.* 27, 274) and Speckled Yellow *Pseudopanthera macularia* (I. and B. Wallace, a small colony in College Valley 28. 5.1974).

From the North Tyne valley and Kielder area have come records of the Golden Rod Brindle *Lithomoia solidaginis*, Sand Dart *Euxoa cursoria*, Least Yellow Underwing *Noctua interjecta*, Straw Point *Rivula sericealis*, Large Blood Vein *Cyclophora punctaria*, Sharp – angled Carpet *Euphyia unangulata*, Clouded Silver *Lomographa temerata* (also from near Kyloe), Scorched Wing *Plagodis dolabraria*, London Brindled Beauty *Licia hirtaria* and Oak Beauty *Biston strataria*.

From the south and west of the County some of the more notable recent records are the Satin Beauty *Deileptenia ribeata* (Dipton Woods, 28.7.1974 J. D. Parrack), Large Seraphim *Lobophora halterata* (Staward 3.6.1977, J. D. Parrack), Barred Carpet *Perizoma taeniatum* (Oakpool, 5.8.1975, D. A. Sheppard) and Pretty Pinion *Perizoma blandiata* (Bedlington, 12.7.1977, J. D. Parrack).

Of species which have been confused mention should be made of Lempke's Gold Spot *Plusia putmani* ssp. *gracilis* formerly confused under *P. festucae*. Both occur in Northumberland and the evidence suggests that *gracilis* is the more common and widespread of the two. According to Baron de Worms the general food-plant of *gracilis* is not known though it is thought that it may be Iris like that of *festucae* (*Ent. Gaz.* (1978) Vol. 29, 26). The Marsh Square Spot *Diarsia florida* was formerly confused with the Small Square Spot *D. rubi*. Both are present in the County e.g. both occur at Prestwick Carr but northwards *rubi* seems to decrease and I never took it in Berwickshire though *florida* was widespread and fairly common. The distribution of *rubi* in VC 68 needs to be ascertained by light trapping in May and September as this species is double brooded unlike *florida* which flies in June and July.

The three common species of Ear moths all occur in Northumberland though their distribution is not adequately worked out. Their identification is only ascertained for certain by examination of the genitalia. Similarly all dark specimens of the Grey Dagger should be kept for examination of the genitalia. We still do not have an authentic record of the Dark Dagger *A. tridens*. Of species which are apparently increasing the Wall butterfly is of interest as it is now present again in the Tweed area after a long period of eclipse. The range of the Scotch Brown Argus *Aricia artaxerxes* especially at the coast should also be investigated while the Scotch Argus *Erebia aethiops* long ago recorded at Elsdon and Fawdon in the Cheviot area may still be awaiting rediscovery in some area of marginal ground where its food-plant *Molinia caerulea* grows. It would also be worthwhile searching the Langleeford valley and Ingram valley to confirm the old records of the Beautiful Brocade and Triple-spotted Clay. The same is true for John Finlay's record of the Copper Underwing now known to be represented in Britain by two species *Amphipyra pyramidea* and *A. berbera*. Again we would like to know if the Speckled Wood butterfly is still present in the County. We have no certain records for this century. The Green Hairstreak is well known in south Northumberland but has not been taken in VC68. It should be sought for in May at localities having plenty of Bilberry. These examples must suffice to show that much remains to be done to complete our knowledge of the habits and distribution of the Lepidoptera of Northumberland.

(c) **Some Berwickshire Lepidoptera Records, 1952–1954.**

I came to Scotland from North Staffordshire in 1945 but it was not until early 1952 that I resumed systematic moth collecting. As a boy I collected in and around my home town of Todmorden on the border of Yorkshire and Lancashire, my favourite locality being Hardcastle Crags near Hebden Bridge. My Headmaster at the Todmorden Grammar School Mr A. Radway Allen M.A., B.Sc., gave me much encouragement and identified my specimens. Boys make good collectors and I remember the thrill of making captures new to my mentor. Here in Berwickshire I discovered I was virtually a lone hand, most of the collecting having been done by a few keen collectors last century, notably Simpson Buglass of Ayton, Wm. Shaw of Eyemouth, Andrew Kelly of Lauder, Robert Renton of Fans near Earlston, John and Adam Anderson of Preston near Duns and George Bolam of Berwick-upon-Tweed. Bolam compiled a list of the Lepidoptera of the eastern Borders and published it in the *History of the Berwickshire Naturalists' Club* for 1925–1927 Vols. XXV–XXVI. This list has proved most valuable for comparison and for assessing the status of my own captures. Of course there are many species listed by Bolam which so far I have never seen.

On the other hand there are some which are new or have changed their status. For example, when Bolam compiled his list he had only five records of *Deilephila elpenor* the Large Elephant Hawk moth over a period of nearly a century the first record being that of P.J. Selby of Twizel (Northumberland) in 1837. Now this moth turns up each year and I have records from all over the county.

It is my suspicion that parts of Berwickshire, notably the glens, have never been worked thoroughly, probably due to difficulties of transport in the old days. Thus Bolam had no Berwickshire record of *Drepana lacertinaria*, Scalloped Hook-tip though he knew it to be widely distributed in Northumberland. From my own collecting I can say that it is also widely distributed in Berwickshire. I have collected larvae readily by beating birches at several localities.

Bolam mentions that *Pheosia gnoma* Fab. the Lesser Swallow Prominent is scarcer than *P. tremula* Cl., the Swallow Prominent, but I have found that in the glens and valleys south of the Lammermuirs where Birch is more common than Poplar and Aspen, *gnoma* is easier to obtain than *tremula*. Thus when working a Tilley paraffin lamp this season (1.8.1954) at Kyles Hill near Greenlaw Moor I took eleven specimens of *gnoma*. Nevertheless *tremula* is widely distributed and I have found its larva several times in different localities.

In the village of Gavinton where I live there are twelve mercury vapour street lamps which were erected in the autumn of 1951. With the aid of a net on a nine foot cane I have taken many interesting species at these lamps. One of these is *Drymonia ruficornis* Hufn. the Lunar Marbled Brown for which Bolam had no Berwickshire records. In 1952 I took five in the first week of May. In 1953 I had nine, also in the first week of May and this year I took seven from 12–29th May. Usually they alight on the wooden lamp standard and sit quite still about a foot below the light.

Pterostoma palpina Cl. the Pale Prominent I have found as larvae on Sallows and Poplars in four localities, the best place being Duns Castle Woods in late July. I have found the larvae easy to rear in air-tight honey jars.

Bolam had only one record of *Apatele leporina* L. The Miller in Berwickshire. It was taken on Duns Law in 1873 by a Duns cobbler who collected moths. In 1953 I found two larvae at Gordon Moss, two at Kyles Hill and one at Lees Cleugh, all by beating birch. Unfortunately three died when fully grown, from some unknown cause. This year I found twelve larvae at Gordon Moss and all have pupated successfully.

Amathes castanea Esp. the Neglected or Grey Rustic is a species which Bolam recorded as not common. I found larvae in 1952, and again this year by searching heather in the evenings in May, and successfully reared the moth. A better method of obtaining it, however, is to walk among heather at night with a Tilley lamp when the moths are readily attracted or seen at rest on the heather. In this way I took six specimens on 26th August 1954, including a pair *in cop*.

One of the 'new' Berwickshire species which has come to the village lamps is *Axylia putris* L. the Flame Rustic, though it was known to occur in Northumberland and East Lothian. I took six in 1952, and three in 1953. *Tholera popularis* Fab. the Feathered Gothic, has also come to light, three in 1952, three again in 1953, and one to my Tilley lamp this year. It was taken by Buglass at Ayton Castle in 1876.

Another good record from the village lamps was *Aporophyla lutulenta* Schf. the Deep Brown Dart on 5th September 1953, (not new). Three specimens of *A. nigra* the Black Rustic also came to these lights in the autumn of 1953 and this year I have taken two more.

In the early hours of 27th August 1954 I was walking along the edge of Greenlaw Moor at Kyles Hill. At the time a rough west wind was blowing, but I was in the lee of the hill. I had my Tilley storm lantern lit and suddenly a moth dropped into the heather where it sat quivering its wings about two yards in front of me. I boxed it and discovered it was *Lithomoia solidaginis* Hb., the Golden Rod Brindle. In early September I took two more by sugaring fence posts at the edge of the same moor. So far I can find no records of this species for the Eastern Borders. Its occurrence in Berwickshire is probably because of indigenous breeding but in the light of recent records in S.E. England the possibility of migration cannot be ruled out.

Another similar record is that of *Eurois occulta* the Great Brocade. On the night of 24th August 1954 I placed an electric lamp and white cloth on the village green when a rough easterly wind was blowing. To my surprise I found a good specimen of occulta the Great Brocade sitting quietly on the cloth. The following night I sugared telegraph poles near the village and took another. Bolam says it has been "taken over a wide area at long intervals and singly". These records support the suggestion put forward by Dr Birkett that *occulta* has migrated southwards this season (*Ent. Rec.* 66, p. 240).

In July of this year (1954) a friend sent me some larvae of *Hydraecia petasitis* Dbld. the Butterbur Moth from Yorkshire. Since the butterbur plant grows abundantly by the Langton Burn only a few hundred yards from my home I forthwith took a garden fork and botanical vasculum and proceeded to the nearest likely spot to search for larvae. The first plant which I dug was obviously healthy and contained no grub, so I cast my eyes around and spotted a plant with a mottled yellowish leaf. On digging I found my first Butterbur larva – another new record. Altogether I collected about two dozen from which I reared a good series of moths.

In August this year I went in search of pupae of *Nonagria typhae* Thunb. the Bulrush Wainscot. Wearing wellington boots I waded into the margin of Duns Castle Lake and soon found Reed-Mace plants which had been tunnelled. These were most easily recognized by their failure to form a healthy spike of flowers. Altogether I collected a dozen pupae. Last year (1953) I took a male of this species at a Gavinton lamp about two miles from Duns Castle where the nearest known *Typha* grows. Another related species *Rhizedra lutosa* Hb. the Large Wainscot, also came to street lamps, one in Duns and two in Gavinton.

On 13th August 1952 I took a moth sitting on a beech tree trunk at Kyles Hill which I thought at the time was a variety of *Apamea secalis* L. the Common Rustic. Fortunately I kept it, for on later examination it proved to be *Parastichtis suspecta* Hb. the Suspected. This year I took three more by sugaring near the same locality in August.

On a wet and windy night, 15th September 1952, I sugared telegraph poles between Gavinton and Duns. The last but one patch was opposite a lime tree and on it I took a specimen of *Tiliacea citrago* L. the Orange Sallow, another new record. This year, again at mid-September I took three more at Nisbet, near Duns.

Perhaps the most interesting moth taken this autumn is *Cirrhia gilvago* Schf, the Dusky Lemon Sallow. Bolam did not include this in his list though it has been taken once in recent years at Edrom (by Lieut.-Col. W. M. Logan-Home). I have taken eight at sugar and three at light. All were in the vicinity of this village except for one at Kyles Hill. It is not surprising that this species should occur in Berwickshire as there are many elms and this year there was a great crop of fruits owing to good pollination weather in early spring. Another interesting capture at our village lamps was *Pyrrhia umbra* Hufn. the Bordered Sallow on 28th May 1953. This is a coastal species as the food-plant Rest Harrow grows mostly near the sea in this part of the world. *Zanclognatha tarsipennalis* Tr. the Fanfoot and *Z. grisealis* Schf. the Small Fanfoot are both new records taken more than once.

This year while beating birches near a wood called 'the Bell' between Cranshaws village and the 'Hungry Snout' in the Lammermuirs I disturbed a moth which I netted and discovered it to be *Colostygia olivata* Schf. the Beech Green Carpet. Although well distributed in Northumberland, Bolam had no records for Berwickshire. Its foodplant

Lady's Bedstraw grows commonly about Cranshaws. I find, however, that the place of capture is about fifty yards within the East Lothian Boundary.

Venusia cambrica Curt. the Welsh Wave I have found to be fairly common at Lees Cleugh and Kyles Hill. I have found it by day on tree trunks in the vicinity of Rowans and at night it has come to light along with *Lycometra ocellata* L. the Purple Bar. Four specimens of *Dyscia fagaria* Thun. the Grey Scalloped Bar, were taken this season. Two females were caught flying over heather by day and two males came to my Tilley lamp at night. I also reared a moth from a larva taken on heather in May near the Big Dirrington.

Bolam records *Deuteronomos alniaria* L. the Canary Shouldered Thorn as rare, but each year some have come to the village lamps. *Selenia lunaria* Schf. the Lunar Thorn also came well to light in 1952 (six specimens) and *Colotois pennaria* L. the Feathered Thorn has also occurred frequently.

A strange omission from Bolam's list for these Scottish Border counties in the East is *Lithina chlorosata* Scop. the Brown Silver-line. I have found it to be fairly common locally.

On 17th April 1954 I was surprised to find a specimen of *Ectropis bistortata* Goze in one of my breeding cages. Unfortunately I never kept a record of the locality in Berwickshire from which the larva came. Bolam had no Scottish record for the Eastern Borders.

One "new" species which puzzles me in *Alcis jubata* Thun. the Dotted Carpet. I have taken this moth from several well-separated localities (Gordon Moss, Lees Cleugh, Whiteadder near Cockburn Law and others). I cannot understand how it was missed by the old collectors. This year I reared a series from larvae found in June among lichens on tree trunks and branches. The moth seems more common than *Cleorodes lichenaria* Hufn. the Brussels Lace which Bolam described as "not uncommon in many places especially north of the Border".

Lastly I must mention *Diarsia florida*, Schmidt, the Marsh Square Spot. I have three specimens which I think belong to this species though their identification requires confirmation. They were all taken at light.

In conclusion I may add that Berwickshire has proved a very interesting county in which to collect. It has a good variety of habitats, moors, heaths, glens, mosses, a rocky coastline, and the well wooded plain of the Merse. One can therefore expect a good representation of our northern Lepidoptera.

(d) **Collecting in Berwickshire. January to June 1955.** From *Ent. Record Vol. 67, pp. 254–258.*

During the winter of 1954 – 1955 I made several attempts at pupa digging, chiefly under oaks at Gavinton, Kyles Hill, Aiky Wood, near the Duns to Grantshouse Road and at the Retreat near Abbey St. Bathans. I obtained about two dozen sound pupae, most of which produced *Orthosia stabilis* the Common Quaker and *Q. incerta* Hufn. the Clouded Drab. The more interesting finds were *Erannis leucophaearia* Schf. the Spring Usher from Gavinton and the Retreat, *Orthosia cruda* Schf. the Small Quaker from Aiky Wood, *Bena fagana* Fab. the Green Silver Lines from among fallen oak leaves at Kyles Hill and *Panolis flammea* Schf. the Pine Beauty from under a Scots Pine at Kyles Hill.

During these winter months I also prepared some mercury vapour light traps. One for home use I made out of a tea chest. Two others for taking afield I constructed out of spare bee-hive brood boxes. Each inverted cone was made from four triangular pieces of galvanised sheet iron (gauge 22) fixed together by Meccano nuts and bolts. A mercury vapour light socket was similarly fastened upon the top of each inverted cone. These portable traps are bottomless and when using them I set them down on a white sheet, operating the lamps from a small Pioneer generator in the back of the van. I work a 125- watt and 80-watt lamp simultaneously usually about 200 yards apart. My results seem to show that the situation of the lamp is more important than its power as sometimes I have caught more moths at the 80–watt lamp. Possibly three 80–watt lamps well separated would give the best results.

The first moths of the year were *Phigalia pedaria* Fab. the Pale Brindled Beauty and *Theria rupicapraria* Schf. the Early Moth on 29th January at Gavinton lamps. Two male *E. leucophaearia* Spring Usher hatched from pupae on 2nd and 19th February. Towards the end of February male and female *P. pedaria* emerged in my breeding cages and I obtained a pairing. The female started laying on 1st March and the larvae started to hatch on 1st May. On this date some eggs of *Poecilocampa populi* the December Moth also hatched. These fed much more slowly than those of *pedaria* and the first spun its cocoon about mid-July.

On 30th and 31st March I tried out my lamps at Kyles Hill road-end. I discovered that *Achlya flavicornis* L. the Yellow Horned was just beginning to emerge. On 3rd April I visited the same spot and about two dozen fresh specimens of this lovely moth came to the lamps. The night was clear with moonlight and there was much audible activity among the Pink-Footed Geese on Hule Moss. A Wood Pigeon

Gladys and her brother Fred at Hule Moss on Greenlaw Moor, Berwickshire, with the Dirrington Hills in the background. The lake is a roosting place for large numbers of Pink-footed Geese in autumn and spring.

was serenading among the pine branches at 10.30 pm. Other moths in flight were *Colostygia multistrigaria* Haw. the Mottled Grey, *Alsophila aescularia* Schf. the March Moth, and the first *Orthosia gothica* L. the Hebrew Character was seen. Truly the season had started.

On 7th April I took my lamps to the same site again, but the night was very different. My diary told me that the moon was full, but no moon was visible, the sky being overcast with thick cloud. At first the moths came well, and then the flight died down and then light rain began to fall. I considered packing up but persevered and suddenly about 9.30 p.m. a large black and white moth plunged onto the sheet. It was *Biston strataria* Hufn., the Oak Beauty unrecorded by Bolam for the Eastern Borders, a lovely male specimen. I had already tried for this moth on two nights (5th and 6th of April) at the Retreat below Abbey St Bathans but without success and now it turned up unexpectedly in a locality where birch predominates. My visits to the Oak woods at the Retreat, however, were not without their reward as I discovered *O. cruda*, the Small Quaker to be not uncommon there. Some came to Tilley lamps as well as to MV lamps. Later I found this species to be near Oxendean Pond in Duns Castle Woods. Bolam considered it to be rather local in the eastern Borders.

On 13th April I set out for Gordon Moss and arrived about 7 p.m. to find the gate padlocked. A notice informed me that a key was obtainable at the village police station so I went along to ask for it. I discovered that the policeman was one who had investigated my Tilley lamp operations in the early hours of one morning last August. I borrowed the key and put down my traps and sheets. The night was cool and starry with a slight westerly breeze. I worked the lamps from 8.15 to 11.15 p.m. and was rewarded with three *Orthosia advena* Schf. the Northern Drab and some *Cerastis rubricosa* Schf. the Red Chestnut, *C. incerta* the Clouded Drab, *O. stabilis* the Common Quaker, *O. gothica* the Hebrew Character, *C. multistrigaria* the Mottled Grey and *A. flavicornis* the Yellow Horned. I pushed the borrowed key through the letter-box of the police station door at midnight and was home by 12.30. It was worth it to get another species new to me.

The anticyclonic weather of April, however, gave us cold clear nights and ground frosts, and these were hopeless for collecting. However, on the 19th I tried both Tilley and MV lamps by the old railway near Gavinton – a good place for wild roses. I expected to get *Earophila badiata* Schf. the Shoulder Stripe but by 11 p.m. I decided to switch off in despair. Immediately afterwards I took three specimens at my two Tilley lamps. On 22nd April I again tried my lamps at Kyles Hill road where 20 to 30 specimens of *Nothopteryx carpinata* Bork. the Early Tooth-striped put in an appearance. The moth is fairly common where Birch abounds.

On 28th April I again visited Gordon Moss in company with Dr Macnicol and Mr Pelham-Clinton of Edinburgh. *Lampropteryx suffumata* Schf. Water Carpet and *Colocasia coryli* L. the Nut-tree Tussock were just beginning to fly, but the most interesting captures were two *Agrotis ipsilon* Hufn. the Dark Sword Grass at Dr Macnicol's lamps.

In early May I went in search of *Ectropis bistortata*, the Engrailed. I knew that this species was in the County as I have reared a single larva from some unrecorded locality. I went to the North Bog near Oxendean in Duns Castle Woods. Here there is a mixed wood, chiefly Birch, Elm, Alder, Pine, and Beech – a haunt of the Pied Flycatcher, Wood Warbler and Heron. By searching many hundreds of birch trunks I discovered one male *bistortata* sitting about eighteen inches above ground level in a small depression of the bark. The following evening (8th May) I searched the same locality again and took three more. Two of these were females and one I kept for eggs. These were bright green in colour like the opening birch leaves. I have reared over 100 pupae, one of which hatched in early August. The species must therefore be normally single brooded in Berwickshire.

On 9th May I took my lamps to Oxendean Pond and netted another male bistortata as it was flying away. I also took a single *Drymonia ruficornis* Hufn., The Lunar Marbled Brown. This was a new locality for this species. Later I took others at Kyles Hill and at the Retreat on the Whiteadder. These records show that it is widely spread in Berwickshire.

My best collecting night in May was the 23rd. This was the first warm night after a fortnight of cool weather. I took my lamps to the Retreat and recorded fifteen species including *D. ruficornis* the Lunar Marbled Brown, *Pheosia gnoma* the Lesser Swallow Prominent, *Selenia lunaria* Schf. the Lunar Thorn, *S. bilunaria* Esp. the Early Thorn, *Lophopteryx capucina* L. the Coxcomb Prominent, *Lithina chlorosata* Scop. the Brown Silver-line, *Biston betularia* L. the Peppered Moth, *Euplexia lucipara* L. the Small Angle Shades and *C. rubricosa* the Red Chestnut. Earlier on the same evening I took a freshly emerged female *Saturnia pavonia* L. the Emperor Moth to Greenlaw Moor and put her in a cage on the Ordnance Survey triangulation pillar at 6 p.m. After thirteen minutes a male came flying directly up wind about two feet above the ground. On reaching the pillar it must have overshot the scent as it turned and circled several times unable to locate the cage which was about four feet above ground level. I therefore put the cage down on the ground and the male quickly found it.

On 24th May I operated my lamps at Kyles Hill and took *D. ruficornis* The Lunar Marbled Brown, *Hadena bombycina* Hufn. the Glaucous Shears and *Apatele rumicis* L. the Knotgrass. Later I took three other specimens of *bombycina* at the same place: they arrived shortly after midnight. I also took *S. lunaria* the Lunar Thorn among the oaks at this locality. When sugaring near Gavinton during the previous autumn I caught two "Thorn" larvae descending the same elm trunk on different nights after dark. These produced imagines of *S. lunaria* the Lunar Thorn. Nevertheless this species seems more partial to Oak woods. At Kyles Hill I was pleased to find three patches of the Oak Fern *Thelypteris dryopteris* (L.) Slosson while looking for a spot to place a Tilley lamp and sheet.

On 28th May I took my lamps to the Bell Wood above Cranshaws in the Lammermuirs; but after sunset the temperature dropped quickly to 32°F. The frost crystals were glistening on my sheets and I had to pack up. Only two moths came to the lamps, *G. rubricosa* the Red Chestnut and *Rusina umbratica* Goze. the Brown Rustic. I saw a fox come out among the heather along the opposite hillside and I saw it make a quick rush after which I heard a rabbit squeal. Myxomatosis had not yet arrived there. The following night I visited Kyles Hill again and took *Eumichtis adusta* Esp. the Dark Brocade. This species had previously evaded me being scarce at the Gavinton lamps where

Hadena thalassina Hufn. the Pale Shouldered Brocade is more common. However, the portable lamps have revealed that *adusta* is quite common on upland ground. I found it later on Penmanshiel Moss near Coldingham Moor.

On Monday 30th May, I decided to go in search of *Euclidimera mi* Cl. on Cockburn Law. In the boggy Alder wood near the Whiteadder I found some lovely fresh *Hydriomena coerulata* Fab. the May Highflier and the first specimen of *Electrophaes corylata* Thun. the Broken Barred Carpet. The smell of dead rabbits was nauseating. On the Law I saw at least six moths flying over the heather, but old Mother Shipton was too fast for me. The first I saw settled and I got my net over it, but it escaped. However, I took a worn female Glaucous Shears which laid over a hundred eggs. These produced a fine healthy looking brood of larvae which fed up well on Hawthorn; but to my disgust they all died when fully grown having apparently contracted some disease as deadly as myxomatosis – and this after I had transported them down to Lancashire and back during my summer holiday. Some of these *bombycina* L. larvae were dark brown others light brown and a few dirty green.

On 1st June a pupil brought a larva of *Setina irrorella* L. the Dew Footman from the coast. There are colonies of this species at Burnmouth and Eyemouth. The larva pupated and produced a female moth on 8th July. Three specimens of *Anticlea derivata* Schf. the Streamer came to my home trap in Gavinton in the first week of June.

The night of 4th June was dull and warm with some rain. I operated my lamps from 10.30 p.m. to 2.30 a.m. at Oxendean Pond. The temperature never dropped below 53°F. and I recorded thirty-four species. These included *Ecliptopera silaceata* Schf. (abundant) the Small Phoenix, one *Triphosa dubitata* L. the Tissue, several *Celama confusalis* H.-S. the Least Black Arches, one *E. bistortata* the Engrailed, one *Pterostoma palpina* Cl. the Pale Prominent also *Drepana falcataria* the Pebble Hook-tip, *Perizoma affinitata* Steph. the Large Rivulet and *Eupithecia tantillaria* Bdv. the Dwarf Pug. Pleased with these results I made a return visit on 13th June, and in a moment of forgetfulness connected my 80-watt lamp to the generator, omitting the choke. When switched on there was a mighty surge and I fused the lamp. Somewhat humiliated I continued operations with one lamp and took a single *Harpyia furcula* Cl. the Sallow Kitten; but heavy rain put a stop to further collecting and I was forced to retire.

During June a good series of *Clostera pigra* Hfn. the Small Chocolate Tip emerged from pupae. These were reared from larvae obtained on

Dwarf Sallows along the railway side at Gordon Moss during the previous August.

On the evening of 15th June I walked out onto the heather moor between Kettleshiel and the Dirringtons. The sun was shining and there was a good flight of *Eupithecia nanata* Hb. the Narrow-winged Pug, *E. satyrata* Hb. the Satyr Pug., *Epirrhoe tristata* L. the Small Argent and Sable, *Xanthorhoe spadicearia* Schf. the Red Twin-spot Carpet, and *Ematurga atomaria* L. the Common Heath. I found a larva of *Dasychira fascelina* L. the Dark Tussock spinning its kenspeckle cocoon among the tips of heather shoots. On this date the eggs of *H. bombycina* hatched.

On 17th June specimens of *Apatele leporina* L. the Miller and *Lasiocampa quercus* L. var. *callunae* the Northern Eggar, emerged in my breeding cages. Two pupae of *A. leporina* L. the Miller which had lain over two winters still declined to hatch although apparently healthy.

Friday 24th June was a record night. I went to Gordon Moss, arriving about 9 p.m. I treacled telegraph poles and fencing posts along the railway which cuts right through the Moss. I had the lamps on from 10.30 p.m. to 2.30 a.m. and recorded a total of fifty species. Never before had I seen moths come to treacle in such numbers, twenty to thirty on each patch, making a total round about a thousand. There must have been over a hundred *Anaplectoides prasina* Schf. the Green Arches in all their first fresh splendour, *hyppa rectilinea* Esp. the Saxon and *A. leporina* the Miller along with *A. rumicis* the Knotgrass, *A. psi* L. the Grey Dagger, *Diarsia brunnea* Schf. the Purple Clay, *D. rubi* View the Small Square Spot, *Amathes c- nigrum* L. the Setaceous Hebrew Character, *Euplexia lucipara* L. the Small Angle Shades and the first *Triphaena pronuba* the Large Yellow Underwing. At light I took the Prominents, *P. gnoma* Fab. the Lesser Swallow Prominent, *N. ziczac* L. the Pebble Prominent, *L. capucina* L. the Coxcomb Prominent, *Lycometra ocellata* L. the Purple Bar and *Xanthorhoe ferrugata* Cl. the Red Twin-spot Carpet (which I had not recorded before in the county). A Sedge-Warbler broke into song at 2.20 a.m. and as dawn broke the sky cleared and one of my sheets became stiff like a board with a slight ground frost. It was a record night for the first half of the year and it gave me a great thrill to see moths crowding the treacle patches in such numbers.

(e) **Collecting in Berwickshire. July–December 1955. From the** *Entomologists' Record, Vol. 68 pp. 8–15.*

Having had good collecting at Gordon Moss on 24th June I made a return visit on 1st July. The evening was clear, cool, and damp with a N.W. wind at first. Later there was a moon and the night was never really dark. Prospects did not seem promising at the start. Somewhat half-heartedly I treacled along the railway side and then put out my two MV light traps. At dusk I took my net and strolled down through the thick birch bushes where the roe deer sometimes lurk and near the railway I casually netted a black and white moth. On boxing it I immediately realized it was something new. It proved to be *Mesoleuca albicillata* L. the Beautiful Carpet. When I consulted Bolam's list I found that he had no record of this species for the Scotch Border Counties, his nearest locality being in Northumberland. This was a good start for an unpromising evening but although I worked all night, I got little else worthy of note. *Xanthorhoe ferrugata* Cl. the Red Twin-spot Carpet and *Pheosia tremula* Cl. the Swallow Prominent came to the lamps. At first I could not understand the presence of *tremula* the Greater Swallow Prominent among birch, so later I searched for poplars and found them at the Earlston end of the Moss. A hedgehog came sniffing round one of my lamps and I had to roll it away. The treacle patches only averaged about eight to ten moths. Indeed from this date the attractions of treacle steadily declined until September. However, there were some beautiful fresh *Diarsia brunnea* Schf. the Purple Clay which were worth keeping.

On Saturday, 2nd July, I should have visited the Isle of May with a party of ornithologists but the sea was too rough to make the crossing. Instead we visited Fast Castle on the Berwickshire coast, but it was impossible to do much more than spy out the land for intended visits in the future. On the following day we had a tremendous thunderstorm and torrential rain; this heralded great heat and humidity and the amazing hot spell which never really broke for two full months. Conditions for night collecting became almost perfect and I could have set out almost every evening had this been otherwise possible. There was a great nectar flow from clover and possibly this accounted for the failure of the treacle to attract moths.

On 4th July I returned again to Gordon Moss. The track down onto the Moss is very rough and overgrown and in attempting to turn the van in readiness for coming home I ran one rear wheel into a ditch. After lifting out the generator and laying one of my sheets under the wheel for greater purchase I still failed to get away. I had visions of being ditched for the night but providentially two sturdy schoolboys, who had seen me pass through the village of West Gordon, came down to watch the moth catching, and with their timely assistance the situation was retrieved and the old van came out onto dry land not much the worse.

After this I treacled and worked my lamps all night and recorded sixty species. Among these I took two *Xanthorhoe designata* Hufn. the Flame Carpet – new to my County list, and *Plusia festucae* L. the Gold Spot, *Leucania comma* L. the Shoulder Striped Wainscot, *Amathes triangulum* Schf. the Double Square Spot, *Hyppa rectilinea* Esp. the Saxon, *Apatele leporina* L. the Miller, *Cleorodes lichenaria* Hufn. the Brussels Lace and dozens of *Anaplectoides prasina* Schf. the Green Arches. *Venusia cambrica* Curt the Welsh Wave came to light again and it would therefore seem likely that birch is its food plant in this locality. A single *Phalaena typica*. the Gothic was taken, a species that seems strangely uncommon in some of these more natural habitats.

Throughout the season I worked a home light-trap at Gavinton and it was interesting to return home after dawn and see what had been flying about one's doorstep. On 7th July *Pyrria umbra* Hufn. – the Bordered Sallow appeared at the trap. There was also a swarm of small chocolate and white caddis flies which I identified as *Leptocerus albifrons* L.

On 8th July a sea mist came in but this did not stop the moths from flying. Forty-eight species came to my home trap including *Cleorodes lichenaria* Hufn. the Brussels Lace and *Dyscia fagaria* Thun. the Grey Scalloped Bar. The latter must have travelled one to two miles from the nearest heather.

The 10th July I treacled along the Greenlaw Road beyond Polwarth and operated lamps at Kyles Hill road-end. I netted a nice *Thyatira batis* L., the Peach Blossom flying along the roadside hedge at dusk. A treacle patch on a Scots Pine trunk produced a perfect female specimen of *Eurois occulta* L. the Great Brocade. This was the first and best of a long series taken this season. Altogether I recorded thirty-nine specimens of this insect between 10th July and 30th August from six well separated localities viz. Kyles Hill, Gavinton, Oxendean Pond, Spottiswoode, Retreat and Bell Wood. These records suggest that the species is more common and widely established in Berwickshire than was formerly supposed. I noticed that many of these specimens only came to my lamps late in the night. The following night I made a return visit to Kyles Hill but results were disappointing. However, as dawn was breaking, about 2.40 a.m. I heard a quail calling in the fields between Polwarth and Kyles Hill.

Between 13th and 20th July I was in the West Riding of Yorkshire and re-visited former collecting haunts near Hebden Bridge. The Hebden Water below the little Baptist Chapel at Blake Dean was still as attractive as ever, the stream still tumbling over the boulders of millstone grit, graceful clumps of Mountain Shield Fern (*Thelyptris oreopteris*) growing on the steep slopes and here and there a large *Aeshna* dragonfly hawking flies.

I motored back to Scotland on 20th July and was at Gordon Moss again on the 21st. After putting down my lamps and flex I was returning to the van just at dusk when I heard a strange noise. I thought that a bird must have flown into the van, but I could find nothing. Soon after, about 10.30 p.m., I discovered the cause when I saw a large water beetle *Dytiscus marginalis* L. circling the van roof like a little helicopter. Suddenly down it came with a smack and slid off the roof onto the bonnet. During the next half hour several more came and I could only guess that the reflection from the van roof must have resembled an inviting pool of water. This goes to prove how much these beetles migrate at night. Later I caught one in a MV trap along with large numbers of a small Water Boatman (*Corixa*). I recorded fifty species of Lepidoptera including *Plusia festucae* L. the Gold Spot, *Plusia bractea* Schf. the Gold Spangle, *Geometra papilionaria* L. the large Emerald (all abundant) and the first *Bombycia viminalis* Fabr. the Minor Shoulder Knot.

On 23rd July fifty-six species came to my home trap including *Xanthorhoe munitata* Hb. the Red Carpet, *Anaitis plagiata* L. the Treble Bar and *Alcis jubata* Thun. the Dotted Carpet.

Sunday 24th July, was a beautiful sunny day and I visited Jordon Law Moss near Spottiswoode – the former home of Lady John Scott who wrote the music to "Annie Laurie". I was with a botanist friend and we were pleased to find the Round-leaved Sundew (*Drosera rotundifolia* L.) in flower. I netted two rather worn specimens of *Coenonympha tullia* Mull. the Large Heath butterfly and one *Aphantopus hyperantus* L. the Ringlet butterfly, without any rings. The Forestry Commission has ploughed deep trenches in the peat and planted young pines on the ridges between.

The night of 26th July was clear but the moon soon set and conditions were quite good. I went to Kyles Hill and ran a length of flex from the van down an old track by a dry stone dyke, a distance of about a hundred and fifty yards. I set one of my lamps on the bank, level with the top of the dyke overlooking the heather moor on one hand and the mixed pine and birch wood on the other. Here I took my first specimen of *Trichiura crataegi* L. the Pale Oak Eggar and also *Plusia interrogationis* L. the Scarce Silver Y, *Thera firmata* Hb. the Pine Carpet and *Eupithecia goossensiata* Mab. the Ling Pug.

On 27th July I visited the home of friends at Spottiswoode and worked one of my lamps in their garden and the other just outside. Treacle was a failure – only one moth – *Thyatira batis* L. the Peach

Blossom came to the bait and got caught with its wings. However, the lamps brought in fifty species including *Eurois occulta* L. the Great Brocade and *Apamea furva* Schf. the Confused. The hour before dawn seemed best and as daylight was breaking and the swallows twittering, a nice specimen of *Anaitis plagiata* L. the Treble Bar, appeared. This species seems to be very widely distributed in the county.

On 29th July I visited the Bell Wood, Cranshaws. This was my first visit since 28th May when the temperature had fallen to freezing point. I treacled a long line of telegraph poles by the roadside and noted *Lycopodium clavatum* growing among heather. Myxomatosis had arrived since my last visit and in places the smell was very bad. The night was starry and the temperature fell to 47°F. During the evening there was a good flight of *Hepialus sylvina* L. the Orange Swift. The commonest moth at treacle was *Amathes xanthographa* Schf. the Square Spot Rustic. At light I took several *Colostygia salicata*. the Striped Twin-spot and one *Euxoa tritici* L. the White-line Dart which came just at dawn. Both these species were new to me. Others worthy of note were *Trichiura crataegi* L. the Pale Oak Eggar, *Lampra fimbriata* Schreb. the Broad Bordered Yellow Underwing, *Eurois occulta* L. the Large Brocade, and *Anaitis plagiata* L. the Treble-Bar. I also took a fine series of *Stilbia anomala* Haw. the Anomalous (all males). The Bell Wood is an interesting locality of scrubby birch on Silurian strata. It lies between the 800–900 ft. contours at the foot of Bothwell Hill which rises to over 1,300 ft. It is almost surrounded by heather moors and is well isolated from any similar birch locality.

On 31st July I decided to re-visit the Retreat on the Whiteadder below Abbey St Bathans. After treacling many oak trees and laying down my traps I found myself assisted by several boy scouts from the Merchiston Castle School in Edinburgh. One of the commonest species at light was *Alcis jubata* Thun. the Dotted Carpet. It came in scores and settled on the sheets and herbage all round the traps. I took several *Stilbia anomala* Haw. the Anomalous and *Venusia cambrica* Curt. the Welsh Wave, *Eurois occulta* L. the Great Brocade, *Plusia interrogationis* L. the Scarce Silver Y, *Phalera bucephala* L. the Buff Tip, *Pheosia gnoma* Fab. the Lesser Swallow Prominent, *Procus literosa* L. the Rosy Minor. *Geometra papilionaria* L. the Large Emerald, *Eupithecia pulchellata*., the Foxglove Pug and *Thera firmata* Hb. the Pine Carpet. In my home trap I took *Eurois occulta* the Great Brocade and *Apamea furva* Schf. the Confused.

August saw a continuation of the fine weather, but the drought began to have its effect. On 22nd August I was at Gordon Moss again. Treacle was almost a failure attracting only four common species in small numbers. At light *Bombycia viminalis* Fab. the Minor Shoulder Knot, was very common. The number of *Geometra papilionaria* L. the Large Emerald was down to eight whereas on my last visit (21st July) there were between thirty and forty at my lamps. I took the first Ear moths for the season and these proved to be *Hydraecia oculea* L. the Common Ear, also *Parastichtis suspecta* Hb. the Suspected, *Cirrhia icteritia* Hufn. the Common Sallow, *Celaena haworthii* Curt. Haworth's Minor, *Sterrha biselata* Hufn. the Small Fan-footed Wave and *Eupithecia tenuiata* Hb. the Slender Pug. Just before dawn one specimen of *Hydraecia petasitis* Dbld. the Butterbur Moth appeared in one of the traps; I was unaware of the presence of its foodplant and still have not seen it at this locality. During the round of my treacle patches I saw and heard three birds feeding among burr-reed in a wide ditch known to the local people as "the stank". I shone my torch on them and judged that they were Water Rails. They were lighter coloured than Moorhens, white under the tail, skulked with the head down, and made a very loud sharp alarm call.

On 3rd August I visited Penmanshiel Moss in daytime. Around the peat holes two species of dragon flies were very active. I caught a few and later identified them as *Aeshna juncea* L. and *Sympetrum danae* Sulz. Two larvae of *Cerura vinula* L. were found on Sallows and we disturbed a Short-eared Owl.

On 4th August I re-visited the Bell Wood above Cranshaws. During the evening I received a courteous call from the water bailiff who was readily persuaded that my apparatus was intended to catch moths and not trout. At first the night was dark with some cloud but later it cleared and the full moon made it almost as light as day. The temperature dropped to 44°F. so I packed up before dawn after recording forty-six species. Among these was one *Triphaena orbona* Hufn. the Lesser Yellow Underwing, *Xanthorhoe munitata* Hb. the Red Carpet, *Trichiura crataegi* L. the Pale Oak Eggar, *Hydraecia oculea* L. the Common Ear, *Stilbia anomala* Haw. the Anomalous, *Pheosia gnoma* Fab. the Lesser Swallow Prominent (all at light), and one *Apamea furva* Schf. the Confused at treacle.

Coming home a gasket blew in the cylinder head of my van engine and I was left to run virtually on two cylinders. The old van chugged like a steam-roller and I began to wonder whether I would reach home. Fortunately most of the road was downhill and I managed to get to the top of the Stony Moor approaching Duns in second gear. I left my van outside a garage at about 5.30 a.m. and walked the remaining three miles to Gavinton none the worse for the exercise.

At Gordon Moss on 9th August I again took a single *Hydraecia petasitis* Dbld. the Butterbur, very late at night. The temperature at first dropped to 40°F. and then the sky clouded over and the temperature rose to 51 degrees F. There was a great flight of *Cirrhia icteritia* Hufn. the Common Sallow, with some beautiful pale lemon-yellow forms. Treacle was a failure with only one *Triphaena pronuba* L. the Large Yellow Underwing and one *Amathes xanthographa* Schf. the Square Spot Rustic. However, when putting on my treacle I heard a Marsh Tit. This is the most westerly point in Berwickshire at which I have recorded this bird. It is apparently absent from the rest of Scotland. I almost trod on a rabbit with myxomatosis, the first time I had seen it at Gordon.

During the early part of August things became very dry and doubtless this delayed some emergences. In my home trap I took some *Bombycia viminalis* Fab. the Minor Shoulder Knot for the first time (5th and 11th).

The night of 12th August was pitch dark with low cloud and fine drizzle. I worked my lamps from dusk to dawn at Kyles Hill and recorded sixty-two species. These included four *Eurois occulta* L. the Great Brocade very late in the night, five lovely fresh *Aporophyla lutulenta* Schf. the Deep Brown Dart, several fresh *Amathes castanea* Esp. the Neglected or Grey Rustic all of the grey form, *Bombycia viminalis* Fab. the Minor Shoulder Knot, *Plusia festucae* L. the Gold Spot, *Diarsia dahlii* Hb. the Barred Chestnut, *Stilbia anomala* Haw. the Anomalous, *Xanthorhoe munitata* Hb. the Red Carpet, *Agrotis ipsilon* Hufn. the Dark Swordgrass and *Lithomoia solidaginis* Hb. the Golden Rod Brindle. Curlews and Lapwings heralded the dawn which came very gradually. In my home trap I discovered moths in tremendous numbers including *Eurois occulta* L. the Great Brocade, *Thera firmata* Hb. the Pine Carpet and *Venusia cambrica* Curt. the Welsh Wave together with a plague of sexton beetles.

Pleased with these results at Kyles Hill I visited the same place again the following night, 13th August. Again conditions were excellent – dark, warm and calm, with the temperature never below 52°F. As I was putting out my lamps a roe deer bounded away into the Pine wood while a large *Aeshna* dragonfly was working to and fro along the dry stone dyke at the edge of the moor catching its supper in the declining rays of the sun. Treacle was an absolute failure but the lamps brought in more *Lithomoia solidaginis* Hb. the Golden Rod Brindle and *Aporophyla lutulenta* Schf. the Deep Brown Dart, lovely fresh specimens, one of which had sooty black underwings. I also took *Geometra papilionaria* L. the Large Emerald, *Trichiura crataegi* L. the Pale Oak Eggar and *Amathes glareosa* Esp. the Autumnal Rustic. On this date I began to notice that certain species were producing a second brood eg. *Apatele rumicis* L. the Knotgrass, *Ecliptopera silaceata* Schf. the Small Phoenix and *Mamestra brassicae* L. the Cabbage Moth.

On 18th August I searched the vicinity of Gavinton for pupae of *Gortyna flavago* Schf. the Frosted Orange and found nine inside Ragwort stems. I noticed that the larva makes an exit hole for the moth but leaves the epidermis intact so that the "hole" is not obvious from the exterior.

On 19th August I revisited Kyles Hill at night but the only additional species of note was *Amathes agathina* Dup. the Heath Rustic. I took a good series of *Lithomoia solidaginis* Hb. the Golden Rod Brindle and now feel sure that it must be indigenous at this locality. Returning home at dawn I paused to admire the view across the Merse, a view which reminds me of the Weald as seen from the top of the Sussex Downs. Fifteen miles to the south-east lay the mouth of the Tweed and the North Sea glinting in the first rays of the rising sun. From this hill at night one can see the Holy Island light flashing off the Northumbrian coast. The hill itself, like the Dirringtons, consists of a reddish porphyry – an intrusive igneous rock which has withstood denudation better than the surrounding old red sandstone and hence its superior elevation at the present day, commanding a view across the Merse to the Cheviots.

On 20th August a single *Lithomoia solidaginis* Hb. the Golden Rod Brindle appeared in my trap at Gavinton and another on the 25th. I also recorded second brood specimens of *Notodonta dromedarius* L. the Iron Prominent both at Gavinton (21st August) and Gordon Moss (26th August).

Rain fell on 21st August and on the 22nd we had thick sea mist or "haar" while England and Wales sweltered with temperatures in the eighties. At night I went to Duns Castle Lake and operated my lamps from 9.30 p.m. to 3.30 a.m. I treacled many trees round the lake but may as well have saved both my energy and treacle. It was very dark with the mist and all the trees were dripping wet. During the night some birds of unknown species circled high up over the lamps making harsh calls not unlike gulls. I judged they were terns lost on migration. I took twenty-five species of moths including *Nonagria typhae* Thun. the Bulrush.

The warm weather soon returned and on the night of 23rd August the temperature never fell below 59°F. I took a second brood specimen of *Drepana falcataria* L. the Pebble Hook-tip along with *Atethmia xerampelina* the Centre Barred Sallow in my Gavinton trap. The Pyralid *Nomophila noctuella* Schiff. also came to light at Gavinton and Kyles Hill.

On 14th August I visited Pease Bay and walked down the coast towards Siccar point – a spot made famous by the visit of James Hutton the geologist who wrote the "Theory of the Earth" in 1795. There were many *Satyrus semele* L. the Grayling, on the wing and I was pleased to find the Hemlock Water Dropwort (*Oenanthe crocata* L.). At night there was another great flight of moths. I found four *Eurois occulta* the Great Brocade in my home trap, one second brood specimen of *Pterostoma palpina* Cl. the Pale Prominent and one *Deuteronomos alniaria* L. the Canary Shouldered Thorn. The most interesting catch, however, was one *Deuteronomos erosaria* Schf. the September Thorn. This was new to me though the species was taken last century by John Anderson at Preston. Later I took three more at the Retreat.

The following night (25th August) was equally good and I took a specimen of *Nonagria typhae* Thun. the Bulrush off a lamp standard in Gavinton.

On 26th August I revisited Gordon Moss. Treacle yielded very little. At first the night was warm and there was a good flight of moths – up to about 1 a.m. Then the temperature dropped to about 40°F. and the flight stopped. In the first hour or so there was a great flight of *Deuteronomos alniaria* L. the Canary-shouldered Thorn and *Cirrhia icteritia* Hufn. the Common Sallow. I recorded forty-two species, the most interesting being *Oporinia filigrammaria* H.-S. the Small Autumnal Carpet and *Xanthorhoe designata* Hufn. the Flame Carpet. I also noted the first *Agrochola circellaris* Hufn. the Brick.

At Oxendean Pond on 27th August I took two more *Xanthorhoe designata* Hufn. the Flame Carpet. *Eurois occulta* L. the Great Brocade, was still in evidence and I was able to get eggs from fertile females. The following night I took an *occulta* at Gavinton street lamps. The last specimen of the season came to my home trap on 30th August and looked quite fresh.

On 3rd September I visited the Retreat and treacled the Oak trees. It was a fine, warm, slightly moonlight night with a S.W. breeze, the sort of night to keep any moth hunter out of his bed. At last the treacle began to pay dividends. The first moth I took was one I had been searching for, *Anchoscelis helvola* L. the Flounced Chestnut – perfect and fresh – but it was the first and the last. I used to get this moth at Todmorden in Yorkshire, but in Berwickshire it had eluded me. However, I was glad to confirm the old records made last century even if I failed to get a series. Three worn specimens of *Triphaena orbona* Hufn. the Lesser Yellow Underwing, and one worn *Mormo maura* L. the Old Lady came to treacle. As usual the Old Lady was near water. The best captures, however, were several *Aporophyla nigra*

Haw. at light (the Black Rustic) along with *Deuteronomos erosaria* Schf. the September Thorn and *Xanthorhoe designata* Hufn. the Flame Carpet, both new locality records. It was a grand night with a total of thirty-two species.

On 7th September we had another warm night after a hot day. There was a half moon and I treacled roadside trees and telegraph poles between Gavinton and Nisbet Bridge over the Blackadder. Results were somewhat disappointing – thirty-seven moths of nine species. Among these I got my first *Peridroma porphyrea* Schf. the Pearly Underwing. Later I took three more.

On 11th September schoolboys brought me a good specimen of *Macroglossum stellatarum* L. the Humming Bird Hawk Moth, found in Gavinton, and on the 16th a schoolgirl brought me a *Herse convolvuli* L. the Convolvulus Hawk moth found in Duns Square. On this date a single *Dasypolia templi* Thun. the Brindled Ochre appeared in my home trap at Gavinton. Later I took eight more at Gordon Moss (23rd).

On 18th September I tried my lamps at Elba on the Whiteadder and treacled Oak trees. I recorded twenty-three species including *Axylia putris* L. the Flame (at light) and *Aporophyla nigra* Haw. the Black Rustic (at treacle). A specimen of *Tileacea citrago* L. the Orange Sallow came to my home trap.

The season was now drawing to a close but still held some surprises. Between 20th and 23rd September I took four *Omphaloscelis lunosa* Haw. the Lunar Underwing, a species new to me. Friday 23rd September saw me at Gordon Moss again. Treacle was almost a failure, most patches being blank. The temperature fell to 36°F. after sunset but after midnight it rose to 46°F. Between 8 p.m. and 9 p.m.. there was a great flight of *Deuteronomos alniaria* L. the Canary Shouldered Thorn. I counted at least 38. The next most abundant moth was *Citria lutea* Strom. the Pink Barred Sallow and I got a good series of *Oporinia autumnata* Bork. the Autumnal Moth by walking about with a Tilley lamp in one hand and a net in the other. The most interesting catch was *Dasypolia templi* Thun. the Brindled Ochre – eight specimens at intervals all through the night. I also took one *Agrochola lota* Cl. the Red Line Quaker shortly before switching off at 5.10 a.m. after running the generator for nine hours. This proved my last of twelve visits to Gordon Moss this season.

On 7th October I went to Oxendean Pond and lit my lamps about 6.45 p.m. I ran the generator for six hours. At first it was very dark but a half moon rose later. A fox came and barked on the other side of the pond. Later an otter came up the burn from the Whiteadder calling all the time. I heard it plunge into the pond and scatter the wild ducks. Later it worked down the roadside where there are tall thick Sallows. I shone my torch on it and its eyes glowed pink. It made a low growl and disappeared up the path towards the pond. Twenty-two species of moths came to the lamps, including a nice series of *Colotois pennaria* L. the Feathered Thorn and *Chloroclysta siterata* Hufn. the Red Green Carpet along with some very dark banded forms of *Thera obeliscata* Hb. the Grey Pine Carpet.

My last excursion using the MV lamps was made on 11th October to Kyles Hill. Sixteen species were recorded including *Peocilocampa populi* L. the December moth, both sexes and *Peridroma porphyrea* Schf. the Pearly Underwing.

During the last half of October I went over my collection of Ear moths and prepared genitalia mounts. I found three species present. The most numerous species in my collection proved to be *Hydraecia lucens* Freyer, the Large Ear which I had taken at Kyles Hill, Gavinton and Duns Castle. *Hydraecia oculea* L. the Common Ear is very widely distributed as I have recorded it from Gordon Moss, the Retreat, Bell Wood and Gavinton. *Hydraecia crinanensis* Burr. the Crinan Ear, appears more local and scarce as I have only had it from Gordon Moss and Kyles Hill.

The month of November was also marked by spells of very mild weather and numbers of *Phlogophora meticulosa* L. the Angle Shades, came to street lamps. Thus on 12th November I counted eleven at rest near lamp standards in Duns. The last one I saw was on 2nd December.

One new record which was kindly communicated to me by Dr D. A. B. Macnicol and Mr E.C. Pelham-Clinton of Edinburgh was *Calocalpe undulata* L. the Scallop Shell, two specimens being taken at Gordon Moss on 18th July. I find now that the County list of "macros" which I have prepared from all sources known to me stands at 391 species of which I have been able to collect 290 species. Of the latter, 20 species were not recorded by the old collectors.

(f) **Collecting in Berwickshire: January to June 1956.** *Entomologists' Record vol. 68, pp. 206–211.*

During the early months of 1956 I made several excursions to search Sallows for borings of *Sphecia bembeciformis* Hubn. the Lunar Hornet Clearwing. At Kyles Hill, Gordon Moss, Kaysmuir, Duke's Wood, Middlethird Bog and Threepwood, near Lauder (Roxburghshire), I discovered many old borings, but only at one locality did I find living larvae, betrayed by the presence of newly formed frass. This was by the Duns to Greenlaw road, about a quarter mile west of Woodheads Farm.

In March I sawed the Sallow trunk and removed about one foot, which I placed on a tray of damp sand in an old meat safe. The larvae continued to produce frass until June, when I discovered that three larvae had left their borings and were lying in the bottom of their cage looking very shrunken and undernourished. I concluded that the wood had become too dry, and though I replaced the larvae in their borings I failed to get any moths. Perhaps it would have been better to have left the sawing of the Sallow trunk until late May.

I saw the first moth of the year fluttering round a Gavinton lamp on 29th January, but failed to catch it. On 6th February eight *Phigalia pedaria* Fab. the Pale Brindled Beauty and four *Theria rupicapraria* Schf.the Early Moth appeared at these lamps, and others followed through the month.

After severe wintry weather towards the end of February, March came in with some fine spring-like days and on the 3rd I found a larva of *Phragmatobia fuliginosa* L. the Ruby Tiger, sunning itself at Kyles Hill. Others were found on the 11th March at Elba on the banks of the Whiteadder. They spun up within a week but one produced puparia of a Tachinid fly.

On 9th March I had taken *Alsophila aescularia* Schf. the March Moth, *Erranis marginaria* Fab. the Dotted Border and *E. leucophaearia* Schf. Spring Usher, at Gavinton lamps. These street lamps keep one well informed of emergences in the early part of the year when it is scarcely profitable to work a solitary MV light trap. On 24th March I found a young larva of *Dasychira fascelina* L. Dark Tussock moth at Kyles Hill, but as it went into aestivation I released it. I also searched Oak trunks at the same locality and found one *E. leucophaearia* the Spring Usher sitting about two feet above the ground.

On 25th March, I again searched Oak trunks at the Aiky Wood, near Whitegates on the Duns-Grantshouse road, and found another *E. leucophaearia* Spring Usher together with one *P. pedaria* Pale Brindled Beauty and one female *Colostygia multistrigaria* Haw. the Mottled Grey; it was a beautiful sunny morning with Skylarks and Chaffinches singing and several times I heard the 'yaffle' of a Green Woodpecker in the woods bordering the Whiteadder. At night I tried my MV lamps for the first time this season at Kyles Hill road (Bent's Corner). The temperature was at 50°F. to begin with but later fell to 40°F. It was calm and the full moon was obscured by cloud. Between 7.15 and 9.45 p.m. I recorded nine species, including about eighty *Achlya flavicornis* L. the Yellow Horned, three *Orthosia incerta* Hufn. Clouded

Drab, and hibernated specimens of *Eupsilia transversa* Hufn. the Satellite, *Conistra vaccinii* L. the Chestnut and *Xylena exsoleta* L. Cloudy Swordgrass. I was also visited by two Greenlaw policemen who had been given a telephone call by a well meaning motor cyclist who had seen my lamps and imagined vain things concerning their purpose.

On 27th March a pupil brought me a nice batch of larvae of *Setina irrorella* L. the Dew Footman from the sea braes about one mile north west of Burnmouth. Later I visited this spot and found the Dew Footman larvae in fair numbers both on the rocks and among the herbage.

On 31st March I visited Threepwood Moss near Lauder. During last century the late Andrew Kelly, one of several Berwickshire collectors of that time, recorded *Dasychira pudibunda* L. the Pale Tussock from near this locality. So far this species has evaded me and I cannot help wondering whether any collector who reads these words knows of its occurrence in Scotland. As I walked round the moss searching Sallows for *bembeciformis*, Lunar Hornet Clearwing borings, I found larvae of *Lasiocampa quercus* L. var. *callunae* Northern Eggar, and *P. fuliginosa* Ruby Tiger sunning themselves on the bushy heather. A Short-eared Owl was hunting over the heather, gliding and turning on its long wings, which gave it the appearance of a harrier. Curlews and Reed Buntings were back at their breeding haunts and a large dark fox came trotting down a field where some tups were feeding unconcernedly. As I climbed the wire fence to leave the moss I accidentally dislodged a specimen of *A. flavicornis* the Yellow Horned which I had overlooked.

On 1st April I visited Dogden Moss, approaching by way of the Kettleshiel Burn and the Kaimes – large gravel ridges bordering the Moss to the north and supposed to be of glacial origin. My object was to spy out the land and look for larvae of *Macrothylacia rubi* L. the Fox Moth of which only one was found. Golden Plover, Black Grouse, Redshanks and Curlews were seen, and one rabbit. At night I took my lamps to Kyles Hill and worked them near the Oak trees hoping to get *Biston strataria* the Oak Beauty, but it failed to appear. Eight species were recorded including *Orthosia gothica* L. the Hebrew Character.

On 7th April I was at Gordon Moss in the company of Dr Macnicol and Mr Pelham-Clinton of Edinburgh. A south-west wind rather spoiled the moth flight though the temperature remained steady at 44°F. The most interesting species taken was one *Dasypolia templi* Thun. the Brindled Ochre. Not even *O. gothica* the Hebrew Character nor *O. stabilis* the Common Quaker put in an appearance. The following night I took *stabilis* abundantly at Oxendean Pond and a single *Panolis flammea* Schf. the Pine Beauty.

On 9th April I paid my first visit with MV lamps to the Hirsel, near Coldstream (by kind permission of the Earl of Home). I pitched my lamps under the large Oak trees in the valley of the Leet, near Montagu Drive. It was my surmise that the Oak Beauty *B. strataria* might be breeding there and my hopes were more than realized when at least 38 specimens of this moth put in an appearance. There is no doubt therefore that this species is established in Berwickshire. Altogether I recorded 10 species between 8 p.m. and 12.30 a.m. when I switched off, well satisfied with this first visit to a new locality. Motoring home I nearly struck a Barn Owl which rose from the side of the road. The following night 10th April I took one more *strataria* the Oak Beauty at a Gavinton lamp.

On 12th April I was back at Gordon Moss and worked my lamps from 7.45 p.m. to 10.30 p.m. The wind was moderate north-westerly and the temperature steady at 43°F. I recorded eight species, including thirteen fresh *Orthosia advena* the Northern Drab and one female *D. templi* the Brindled Ochre. I was also surprised to find one *Orthosia cruda* the Small Quaker as I know of no Oak trees on the Moss. On this date the first of a good series of *Ectropis bistortata* Goze the Engrailed emerged, bred from a female taken in Duns Castle Woods.

On 20th April I revisited the Hirsel, but the night became clear with moonlight and ground frost. No *strataria* came on this occasion but I took *Earophila badiata* the Shoulder Stripe and *Eupithecia abbreviata* the Brindled Pug. I packed up about midnight, the flight having virtually ceased.

Again on 21st April at Gordon Moss the temperature fell rapidly to 30°F. and I had to finish collecting at 10.30 p.m. I saw my first *Nothopteryx carpinata* Bork. the Early Tooth-striped. Back at Gavinton I found the thermometer at 45°F. and a good flight of moths round my garden trap.

On 22nd April I visited Dogden Moss again by day, approaching from Hallyburton Farm near Greenlaw. The only moth seen was a male Emperor *Saturnia pavonia* L. but I noted that the habitat seemed well suited for *Coenonympha tullia* Mull. the Large Heath butterfly and this surmise proved correct as later in the year (12th July) I found this butterfly flying abundantly. I also noted a good growth of Cranberry (*Oxycoccus palustris*) and wondered whether *Carsia paludata* Thun. the Manchester Treble Bar could be found here in August.

I was at Gordon Moss again on 24th April hoping to get *Orthosia gracilis* Schf. the Powdered Quaker but ground frost put an early stop to collecting. *Cerastis rubricosa* Schf. the Red Chestnut appeared in good numbers. Back at Gordon again on 28th April I was rewarded by my first *gracilis* the Powdered Quaker. This species had been recorded from only one other locality in Berwickshire, viz. Pease Dean – by James Hardy, over a hundred years ago. The temperature again dropped below freezing point and I returned home soon after midnight.

My next quarry was *Odontosia carmelita* Esp. the Scarce Prominent and on 4th May I took my lamps to the Birch strip near the main Greenlaw road west of Polwarth. I placed my lamps on the south-west side about 150 yards apart, but results were disappointing – eight common species appeared but no *carmelita*. Last year I tried hard to get this species at Kyles Hill where Birch abounds, but drew a blank, so I began to doubt whether it could be in the county. Bolam recorded it in 1898 at Foulden Hag, but the Birches there have been cut and replaced by conifers. He also recorded a specimen bred from a larva taken at Earlston by Mr Haggart of Galashiels in 1901. These were the only Berwickshire records known to me but sufficient to raise hopes. On 5th May therefore I went to the Hirsel and after seeing the gamekeeper I pitched my lamps on one of the rides in Kincham Wood. This was formerly an Oak wood but within recent years the oaks have been felled and in their place are thickets of self-sown Silver Birch, Ash, Hazel, Crab Apple and Privet. The night was very windy but I was able to find sheltered spots and switched on my lamps about 9.30 p.m. At 10.20 p.m. I found two *Chaonia ruficornis* Hufn. the Lunar Marbled Brown in one trap. Then I walked back to my other trap and saw a moth come into the cone; immediately I recognized the yellow flashes on its wings and knew it was *carmelita* – the Scarce Prominent. This was the only one I took that night, but it was sufficient to show that the species was probably breeding among the young Silver Birches. I also recorded two *O. gracilis* the Powdered Quaker (a new locality), one *Selenia lunaria* Schf., Lunar Thorn (an early date) and several *S. bilunaria* Esp. Early Thorn in a total of thirteen species.

Back at the same place on 7th May I took two more *carmelita* at 9.45 p.m. and 10.15 p.m. (B.S.T.). Again there was a strong S.W. wind which rather spoiled collecting though the temperature stood at 56°F. I returned the following night (8th May) and saw a roe deer near where I put my lights. At first there was a cool wind but later this died down. Another *carmelita* came about 10.50 p.m. and then no more until suddenly four appeared between 11.45 p.m. and midnight. Other species noted were *Pheosia gnoma* Fab. the Lesser Swallow

Prominent, *Ectropis bistortata* Goze the Engrailed, *P. flammea* the Pine Beauty, *Colocasia coryli* L. the Nut Tree Tussock and *Ecliptopera silaceata* Schf. the Small Phoenix.

On 12th May I returned to Kincham Wood in the Hirsel but failed to take any more *carmelita*. *Celama confusalis* H-S, the Least Black Arches, and several *Lithina chlorosata* Scop. Brown Silver-line were present in a total of eighteen species.

On 14th May I went to Gordon Moss. Sedge Warblers and Reed Buntings were singing though the evening was rather cool. Fourteen species came to light including one fresh *Xanthorhoe ferrugata* Cl the Red Twin-spot Carpet, and one early *Hadena thalassina* Hufn. the Pale-shouldered Brocade.

At Kyles Hill on 18th May I worked one lamp in the disused quarry at the edge of the heather moor and the other I placed to the south overlooking the belt of Oak trees. Results were disappointing – many *gothica* Hebrew Characters and *stabilis* Common Quakers and one *Hadena bombycina* Hufn. the Glaucous Shears.

On 19th May I paid a visit by day to a moor by the Hen Toe Bridge near Abbey St. Bathans. Here six Dotterels had been reported. We found them feeding on a bare patch of burnt heather. Afterwards we searched the moor for larvae and found one *Dyscia fagaria* Thun. the Grey Scalloped Bar. My small daughter spotted a cocoon of *Macrothylacia rubi* L. the Fox Moth which later produced a fine female moth and was very useful for assembling. On 21st May I visited Gordon Moss again. The day had been hot but the night was cold with ground frost. Only four species came to my lamps including one *Cerura vinula* L. a male. I caught a fresh *Xanthorhoe designata* the Flame Carpet, under my hat while laying down the flex.

On 23rd May I was back at Kyles Hill Quarry and succeeded in taking three *H. bombycina*, the Glaucous Shears and one *D. fagaria*, like the Grey Scalloped Bar – an early date. Back again on the 25th I failed to get any more *bombycina* but took one female *M. rubi* the Fox Moth. The night became very clear and cool with a full moon.

May 26 was a glorious sunny day and I visited Coldingham and walked down the coast to Linkum Bay. In a little gully near the south end of this bay I netted one specimen of *Cupido minimus* Fues. the Small Blue (an early date); a few *Coenonympha pamphilus* were also on the wing (the Small Heath butterfly).

On 27th May my pupa of *M. rubi* the Fox Moth, produced a fine female moth so I took it to Kyles Hill and set it down in a cage on a grassy knoll behind the quarry. Between 7 p.m. and 8.15 p.m. over twenty males assembled although there was a cool easterly breeze. In the afternoon I saw a few *Anarta myrtilli* L. the Beautiful Yellow Underwing feeding at Bilberry flowers at the same locality.

On 29th May I visited Kincham Wood again, thinking I might possibly get *Drymonia dodonaea* Schf. the Marbled Brown, but in this I was unsuccessful. I recorded thirty-two species, including *Thyatira batis* the Peach Blossom, *Tethea duplaris* L. the Lesser Satin Moth, four fresh *Deilephila elpenor* L. the Large Elephant Hawk, *Pterostoma palpina* Cl. the Pale Prominent, and one *Scoliopteryx libatrix* L. the Herald.

On 3rd June I visited White Gates on the Duns-Grantshouse road in order to beat the juniper bushes growing at the edge of Drakemire. I soon had large numbers of larvae of *Eupithecia sobrinata* Hb. the Juniper Pug though I discovered later that a good proportion were parasitized by a small chalcid. The first imago emerged on 17th July.

On 7th June I visited the Retreat, where I treacled Oak trees and worked the MV lights from 10.30 p.m. to 1.30 a.m. Thirty-one species were recorded, including *Eupithecia pulchellata* Steph. The Foxglove Pug, *T. batis* the Peach Blossom, several *Apatele rumicis* L., the Knot Grass and *A. psi* L. the Grey Dagger. A large fresh female *Biston betularia* L. Peppered Moth, pale form, was taken near a treacle patch.

On 9th June I tried a new locality at Paxton Dean not far from the Tweed. Unfortunately the night became clear and cold (37°F.) and was never really dark. Only eight species appeared including *Laothoe populi* L. the Poplar Hawk and *Agrotis segetum* Schf. the Turnip Moth. I have noticed that *L. populi* males have a very late flight, usually about 2 a.m. while females fly soon after dusk.

I walked up the coast on 10th June from St. Abbs to Pettico Wick, but insects were very scarce. Returning home we stopped on Coldingham Moor about 6.30 p.m. and a good flight of *M. rubi* Fox Moth males and *Eupithecia nanata* Hb. Narrow Winged Pug was in progress.

On 14th June I was at Gordon Moss and treacled telegraph poles and fence poles along the sides of the railway. I had my lamps on from 10.45 p.m. to 2.45 a.m. The night was cloudy, calm but cool (42°F.). Thirteen *L. populi* (all males) came to light and one *Harpyia furcula* Cl. Sallow Kitten. At treacle I took one *S. libatrix* the Herald and three fresh *Hyppa rectilinea* Esp. the Saxon. The total was twenty-five species. The following night I treacled at the Hirsel and worked my lamps near Montague Drive. Another *S. libatrix* the Herald came to treacle and I took a good series of *Xanthorhoe designata* Hufn. the Flame Carpet (about twenty) at light. The total catch was forty species.

On 20th June I visited Broomhouse on the Whiteadder where *Apamea unanimis* Hb. the Small Clouded Brindle was recorded as not uncommon about eighty years ago. I failed to get *unanimis* but took

View across the River Tweed outside Tweedmouth House, near the south-west end of the old road-bridge, Berwick-upon-Tweed. Photograph by Neil Potts, Berwick.

thirty-nine species including another *S. libatrix* the Herald, some fresh *Perizoma affinitata* Steph. the Rivulet and *Zanclognatha grisealis* Schf. the Small Fanfoot. The following night a single *Hadena serena* Schf. the Broad-barred White came to my garden trap. This species was new to me, although Bolam recorded it as common about Berwick-upon-Tweed. On the same night I visited Kyles Hill and took three female *M. rubi* at light in the quarry (Fox Moths). I also got one *Entephria caesiata* Schf. Grey Mountain Carpet which is an early date, one *Hadena nana* Hufn. the Shears and two *H. bombycina* the Glaucous Shears, several *D. fagaria* Thun. the Grey Scalloped Bar and one *Bena fagana* Fab. Green Silver Lines just before dawn. In all, twenty-nine species were taken.

I visited the Bell Wood above Cranshaws on 23rd June; there was low cloud and drizzle. Treacle was well attended and moths came to light in good numbers though everything was wet and it was rather unpleasant climbing up the steep bracken-covered hillside to my 125-watt lamp. I recorded forty-eight species, among which was a very dark *Triphaena orbona* Hufn. Lunar Yellow Underwing. I was also pleased to take a good series of *H. nana*, the Shears which species seems to be rather local in Berwickshire. Other species listed were *Drepana falcataria* L. Pebble Hook-tip, *D. lacertinaria* L. Scalloped Hook-tip, *Phalera bucephala* L. the Buff-tip, *D. fagaria* the Grey

Scalloped Bar, *H. furcula* the Sallow Kitten, and one male *M. rubi* the Fox Moth (at light). I also recorded the 'prominents' *capucina*, the Coxcomb Prominent; *ziczac*, the Pebble Prominent; *gnoma*, the Lesser Swallow Prominent; *dromedarius*, the Iron Prominent; and several *D. fascelina*, the Dark Tussock; one *Eupithecia pulchellata* Steph. the Foxglove Pug and one *S. lunaria* the Lunar Thorn.

On 23rd June a single *Plusia gamma* the Silver Y appeared in my garden trap, and on the 25th I took my first black *B. betularia* the Peppered Moth. Prior to this I was inclined to think that all the *betularia* in Berwickshire were of the typical peppered form as I have seen scores, if not hundreds, of these. Later I took another black specimen at the Hirsel on 29th June. I would estimate, however, that in Berwickshire the black form averages less than one percent of the total population.

I was again at Kyles Hill on 26th June hoping for *Apatele menyanthidis* View. the Light Knotgrass, which however, failed to appear although I took one *Semiothisa liturata* Cl. Tawny Barred Angle and three *H. rectilinea* the Saxon which was a new locality record for the Saxon. *P. palpina* the Pale Prominent and *Venusia cambrica* the Welsh Wave appeared and another *B. fagana* Green Silver Lines was taken just before dawn. I finished at about 2.30 a.m. with a total catch of about thirty species in spite of a cool northerly wind.

On 29th June I re-visited Kincham Wood at the Hirsel. The night was good with a temperature of 54° F and slight drizzle which turned to heavy rain at dawn. Sixty-seven species appeared, including *Hadena conspersa* Schf. the Marbled Coronet which was new to me. Other species noted were *Pheosia tremula* Cl. the Greater Swallow Prominent, *Z. grisealis* the Small Fanfoot, *Cleorodes lichenaria* Hufn. the Brussels Lace, *T. batis* the Peach Blossom and the aforementioned black *betularia*.

On the last night of June I decided to visit the coast. This was my first experience of working the MV lamps within sight and sound of the sea. After visiting the farmer at Fleurs Farm and asking permission, I ran my van down the side of one of his fields and parked overlooking Linkum Bay between Coldingham and Eyemouth. I put one lamp on a little eminence commanding a wide view of the bay and the other I placed on an old path near the foot of the grassy braes. The evening started with rain, but this cleared and I had a great time. I took three species that were new to me, viz. *Deilephila porcellus* L. the Small Elephant Hawk (eight at dusk), *Eupithecia centaureata* Schf. Lime-Speck Pug and *Epirrhoe galiata* Schf. the Galium Carpet. I also took a good series of *H. conspersa* the Marbled Coronet and *Pyrrhia umbra* Hufn.

My daughter Jean on the cliffs near Lumsdaine, Berwickshire, with a view south to Pettico Wick and St Abbs Head.

the Bordered Sallow, *Ortholitha mucronata* Scop. Common Lead Belle, *D. elpenor* the Large Elephant Hawk, *P. palpina* the Pale Prominent, *P. bucephala* the Buff-tip, *Cleorodes lichenaria* Hufn. the Brussels Lace, *Z. grisealis* the Small Fanfoot and *Eupithecia absinthiata* Cl. Wormwood Pug. Dawn came up over the sea very slowly to the cries of Curlew and Herring Gulls, and the night was a fitting conclusion to the first half of the year, with a record catch of seventy-one species.

(g) **Collecting in Berwickshire July to December 1956. From** *Entomologists' Record Vol. 69, pp. 87–91*.

On the evening of 10th August I set out with some misgivings, for Gordon Moss. Heavy dark clouds threatened a thunderstorm, but it failed to materialize. The night was warm and close and what a night for moths. I could have done with sheets the size of a large dining-room carpet. I arrived at the Moss about 8.45 p.m. so late that I had no time to put out treacle. The temperature stood at 58°F and I had to finish before dawn as there were so many moths on the herbage; I felt that I must give them a chance to seek cover before daybreak. I noted seventy-seven species (my highest record so far) but it was not possible to check all that came. I was pleased to find a fine female *O. sambucaria* the Swallowtail Moth showing that its distribution in the county is not confined to the Hirsel. *Mormo maura* L. the Old Lady and *Phalaena typica* L. the Gothic came along with numbers of *Parastichtis suspecta* Hb. the Suspected and I was interested to record *Lygris mellinata* Fab. the Currant Spinach.

On 19th August I visited Cove Harbour and Fast Castle; at Cove I took *P. furuncula* flying by day. The following night I went to Old Cambus Quarry and worked the lamps for seven hours, 9 p.m. to 4 a.m., and treacled fence-posts up the steep south side of the dean. It was a calm night, temperature 53°F and the full moon was well obscured by cloud. When treacling I paused to admire the view northwards beyond the ruins of St. Helen's Church to the Firth of Forth where lights flashed from the lighthouses on the Bass Rock and Isle of May. I took thirty-three species which included several *Euxoa tritici* L., *Amathes glareosa* Esp., *Tholera popularis* Fab., *A. furva* and *Gnophos obscurata* Schf. (carrying small red mites) *Arctia caja* L. again came very late in the night as I had noticed on several previous occasions.

On 22nd August I visited Burnmouth again by day, the weather being ideal. I spent some time netting *G. obscurata* the Annulet, specimens of which were readily disturbed from the rocks and rough herbage. We walked along the cliff tops northwards past the Breeches Rock and Gull Rock to Fancove Head where I netted a variety of *Aglais urticae* L. the Small Tortoiseshell in which half of the right forewing has a white ground colour.

The following night 23rd August, I visited the Hirsel and recorded thirty species at MV lamps. *Tholera popularis* Fab. the Feathered Gothic was very common soon after dark. I also took *Diarsia dahlii* Hb. the Barred Chestnut and one *Cilix glaucata* Scop. the Chinese Character, the latter being evidently a second brood specimen. Many second brood specimens of *Ecliptopera silaceata* Schf. the Small Phoenix were also seen.

On 24th August I was at Kyles Hill from 9 p.m. to 3.30 a.m. Everything was sodden with the recent heavy rain. The first two hours were the most productive and at 1 a.m. I switched off one lamp intending to finish, but as I discovered two *Amathes agathina* Dup. the Heath Rustic, in my second trap which was in the quarry, I decided to continue. Altogether I took thirty-two species including *Oporinia filigrammaria* H-S the Small Autumnal Carpet, *T. crataegi* Fab. the Pale Oak Eggar, *Thera firmata* Hb. the Pine Carpet, *Aporophyla lutulenta* Schf. the Deep Brown Dart, *Bombycia viminalis* Fab. the Minor Shoulder Knot and *Lithomoia solidaginis* Hb. the Golden Rod Brindle. Most specimens were fresh but numbers were small.

On 26 August I went to Burnmouth again and found that a slight landslip had occurred owing to the heavy rain. I treacled fence posts alongside the steep road and for my reward got hundreds of earwigs but no moths. Before dusk I talked with two elderly inhabitants who remembered the visits of Simpson Buglass from Ayton over fifty years ago. Results at the MV lamps were also disappointing as a half moon arose and made the night very light. Only thirty species were recorded, the most interesting being *S. anomola* the Anomalous. Many *A. grossulariata* the Magpie Moth, appeared probably coming from blackthorn bushes. I blamed the wet weather for the paucity of moths. However, more rain was still to come and on 28th August we experienced serious flooding. Between Penmanshiel and Grantshouse the main Edinburgh to Berwick railway was inundated and had to be closed. All over the county burns and rivers burst their banks, reviving memories of the 1948 floods.

On 1st September I visited Old Cambus again. What is normally a dry dean was now in part a long narrow lake. Again the night was quite good but the number of moths was poor. Only eighteen species came to light and treacle, the most interesting being *A. furva* the Confused, *A. lutulenta* the Deep-brown Dart, *Antitype chi* L. the Grey Chi and *Omphaloscelis lunosa* Haw. the Lunar Underwing.

After these rather disheartening conditions things took a surprising turn for the better despite the weather. Thus on 6th September I took *Hadena trifolii* Hufn. the Nutmeg, on a lamp standard in Duns. This was only the second record for the County. The following day brought another surprise as a pupil reported a specimen of *Acherontia atropos*, The Death's Head Hawk sitting out of reach on the Duns Town Hall. We were able to get it with the aid of a borrowed ladder. On the evening of the same day (7th September) I went to the Hirsel and after hurriedly visiting the gamekeepers and treacling a few trees, I set up my light traps alongside the Hirsel Loch. It was a grand night, dull, warm, and fine until dawn. I worked the lamps for nine hours, and to good purpose, the only misfortune being the unnecessary visit of a policeman in the early hours of the morning. The total catch was forty species including two which were new to me viz. *Tholera cespitis* The Hedge Rustic and *Zenobia subtusa* Sch the Olive. I also took twenty lovely fresh *Cirrhia gilvago* Schf. the Dusky Lemon Sallow some of which were very dark specimens. Some fresh *O. lunosa* Lunar Underwing appeared after midnight and two *Deuteronomos erosaria* Schf. the September Thorn, both new locality records. At 2.30 a.m. a female *Herse convolvuli* L. Convolvulus Hawk moth dropped into the grass alongside my 80-watt lamp. It was in a good condition although the scales underneath the abdomen had been rubbed off.

Having taken *atropos* and *convolvuli* within twenty-four hours I realized that an immigration must have occurred. After snatching a few hours sleep on returning home I took my young daughter into Duns and we went round the town searching street lamps. Our search was rewarded when high up on a lamp standard outside the Horn Inn we spotted a large dark moth which could not be identified with certainty. I therefore went to a nearby garage and borrowed a twenty foot ladder and was able to box the moth, which proved to be a most handsome male *convolvuli* in perfect condition.

At night I resolved to try for more and I took my lamps to Kyles Hill but no more were forthcoming. Two *A. ipsilon* appeared, the Dark Sword Grass and also *E. tritici* the White-line Dart, *Aporophyla nigra* Haw. the Black Rustic and *Celaena haworthii* Curt. Haworth's Minor. However, on the following day another *atropos* was brought to me from Grantshouse. The boy who had caught it had killed it with ether and I noticed that it had ejected a pink fluid (meconium) suggesting recent emergence. It had been caught on the 7th. Later on 10th September a little girl brought a third *atropos* to my home. It had been found dead – squashed flat – on the roadside in Gavinton and was dry and brittle as if it had been dead some days. A fourth *atropos* came into my possession during the same week; it was taken on 12th September in the British Legion hut in Duns. No more were seen until October when I received a fifth taken in Ayton on 7th October. It had been found in a potato field and was slightly rubbed. I set it for the gentleman who found it and when I returned it he told me that he had seen two others which were badly worn and were flying and crawling among the potatoes. He also reported one dead larva on the ground, but as this was not kept I was unable to confirm it. Another imago, he said was found nearby in a shop on the main road through Ayton. Later the same gentleman reported that another *atropos* was found on Eyemouth railway station about the end of September. Another gentleman from Coldstream, told me that a specimen was taken in that area in the month of June. Summing up, I myself saw five specimens and received reports of at least five others. I also received one other *convolvuli* from Coldstream and had a reliable report of one at Old Cambus Quarry on 22nd September. This made a total of four *convolvuli* and ten *atropos*, mainly in the period 7th September to 10th October.

On 20th September I worked my lamps by the Hirsel Loch again hoping for *Rhizedra lutosa* Hb. the Large Wainscot but it failed to appear. However the following night I visited Burnmouth and to my surprise one of the first arrivals at the light was a single specimen of *R. lutosa* the Large Wainscot. I was also surprised to see *C. gilvago*

Dusky Lemon Sallow showing how widespread this species is in the County.

During October I twice visited the Aiky Wood near Whitegate with the MV lamps. My first visit on the 12th October was a failure owing to the strong westerly wind, but on 16th October conditions were much better. I placed the lamps near the junipers and treacled roadside trees and fence-posts and oak trees. The lamps were on from 6 p.m. to 9 p.m. but I saw nothing of *Thera juniperata* L. the Juniper Carpet which I was hoping to find. *Poecilocampa populi* L. the December moth came soon after dusk. These were my last excursions with the portable MV lamps this season, and with the advent of petrol rationing it looks as though they will not be used again for some time.

On 20th October I treacled trees and telegraph poles between Penmanshiel and Grantshouse, by the side of the A1 Berwick to Edinburgh road. The patches were all well patronized and I recorded thirteen species including several *Phlogophora meticulosa* L. the Angle Shades and *Xylena exsoleta* L. the Cloudy Swordgrass and single specimens of *A. nigra* the Black Rustic, *Agrochola lota* Cl. the Red-line Quaker and *Agrotis segetum* Schf. the Turnip moth. The most abundant species was *Agrochola circellaris* Hufn. the Brick. During my round I was accosted by the Cockburnspath policeman who stopped his motor cycle to find out what was going on.

Nothing worthy of note was taken during November. One surprising omission from my 1955 records is *Eurois occulta* L. The Great Brocade of which I failed to see a single specimen, whereas in 1955 I recorded thirty-nine. This raises the question as to whether it is mainly an immigrant.

Recently I have brought the County list up to date and find that it now stands at 400 species of "macros" of which I have collected 322 species.

Corrections

In a previous article (*Ent. Rec.* 67, 257) I recorded *Triphosa dubitata* L. at Oxendean Pond on 4 June 1955. Later I showed this specimen to a friend who corrected the identification to *Triphosa cervinalis* Scop. (*certata*) the Scarce Tissue. This agrees better with the date of capture and is the second county record known to me.

Addendum: The Blackneck. *Leigephila pastinum* Treits. Two were taken on August 2 and 6, rather worn, at MV light Burnmouth, probably established on the Wood Vetch that grows on the sea braes, a new record for VC 81.

(h) **Copper Underwings in Yorkshire. From the** *Entomologists' Record, vol 98, 15.9.1986. p. 209.*

Amphipyra pyramidea L. and *A. berbera* ssp. Fletch. — Between 20–28 August 1985 I stayed in Todmorden (SD 9324) West Yorkshire at Todmorden Edge South overlooking Centre Vale Park and Buckley Wood in the valley of the R. Calder. On Wednesday evening 21 August we returned to the guest house a little after 10 p.m. Outside the porch door was an electric light round which a moth was fluttering. I made a grab but missed it and thinking it may have settled in the shade I looked down and saw a moth sitting on the wall with its wings arched over its back. I took it in my hand and soon realized its wings were limp so it must recently have emerged and could not have been the same I had seen fluttering round the lamp. Moreover, it was a Copper Underwing, a species I had never met with or heard of in Todmorden where I lived and collected as a boy in the 1920–40 period. Later this year Dr David A. Sheppard of the Nature Conservancy visited me at Berwick and offered to do a genitalia examination of the moth and confirmed it was *A. pyramidea* male. By a curious coincidence I was in Wooler, Northumberland on 1 September 1985, where I visited Miss Grace A. Elliot at Padgepool House. She showed me a set specimen of a Copper Underwing reared in August from a larva from a hybrid tea-rose in Castle Howard gardens near Malton (SE 7170) N.E. of York during the first week of June. Judging by the underside wing markings depicted in *The Moths and Butterflies of Great Britain and Ireland* p. 154, figs. 6 and 7, I considered the specimens to be *A. berbera* ssp. *svensonii*. This was confirmed by Mr Pelham-Clinton on a visit to the Royal Scottish Museum, Edinburgh.

Porritt had five place records for *A. pyramidea* in Yorkshire and the more recent list in *The Naturalist* 1967–70 added four more, one of which is Triangle nr. Halifax only twelve miles from Todmorden. Is the Copper Underwing extending its range northward or has it been overlooked?

(i) **The Vestal (Rhodomera Sacraria Linn) in Berwickshire.**
From the *Entomologists' Record Vol. 98. P. 122. Notes and Observations.*

On 2 October 1985 during a welcome 'Indian summer' I visited the River Whiteadder between Edrom and West Blanerne (NT8256) to look for fossil plants. Along the side of a stubble field I disturbed a small straw-coloured moth which soon settled again in the stubble. I placed my hand over it and succeeded in boxing it. It was new to me but on arriving home I identified it as the Vestal with characteristic oblique brown marks across the forewings. In the same fields I also saw a Painted Lady *Cynthia cardui* and Silver Y, *Autographa gamma* in flight.

On 14 October I was in a stubble field on the neighbouring farm of Broomhouse Mains again near the Whiteadder. Here I disturbed another Vestal and captured it. Both were given to the Royal Scottish Museum. On 17 October I disturbed and caught a third specimen in rank herbage near the Whiteadder above Hutton Bridge Mill (NT 9254). These captures as far as I know, are the first record for Berwickshire and seem to indicate a considerable immigration. I am now left wondering who was the imaginative lepidopterist who gave this moth its evocative English name?

(j) **The Return of the Orange Tip. From The** *Entomologists' Record, Volume 91, 1979, pp. 16–17, and 42–44.*

(i) *In Berwickshire VC 81.*

The Orange Tip, *Anthocharis cardamines* L. was a fairly well-known butterfly in the Eastern Scottish Borders about the time of the founding of the Berwickshire Naturalists' Club in 1831. Thus in 1832, the founder Dr George Johnston described it as a local species, rare near Berwick but occurring on the road between Paxton and Swinton and also between Swinton Mill and Coldstream (*H.B.N.C.*I., 8).

In 1850 it was seen at Coldingham Moor on 19 June (*H.B.N.C.* III, 5), and in 1880 one was taken by Dr Stuart at Broomdykes (*H.B.N.C.* IX, 295), while others were noted at Humebyres and Gordon Moss by Robert Renton (*H.B.N.C.* IX, 295).

About this time it was also seen in Lauderdale, as recorded by Andrew Kelly in the book *Lauder and Lauderdale* by A. Thomson (1902) though apparently it was becoming scarce, as in 1897 William Shaw wrote "Once common at Gordon Moss but never seen now" (*H.B.N.C.* XVI, 231).

George Bolam, writing in 1925, said that it occurred in the Eyemouth district many years prior to 1887 and also in Duns district, but he had no records for the twentieth century (*H.B.N.C.* XXV, 522). The last recorded year of occurrence in Berwickshire therefore seems to have been 1880.

During the period 1945–66, when I lived in Berwickshire, I never saw or heard of a single specimen in the County and I thus regarded it as probably extinct in the County. In my County list (1957), I wrote, "Is it too much to hope for its rediscovery or is it extinct in the County?" (*H.B.N.C.* XXXIV, 132).

The first known recent occurrence of Orange Tips in Berwickshire, was on the North bank of Tweed below Leaderfoot Bridge on 18 May 1975, when about ten specimens were seen by D.G. Long (*H.B.N.C.* XL, 104); and the following year, a male was seen on *Aubretia* in a garden at Earlston by Henry Polson (A.J. Smith, *Journal Edinb. Nat. Hist. Soc.* 1976, 12).

In 1978, one was seen at Abbey St. Bathans on 24 May 1978 and another at Stichill on the same date (A.G. Buckham). One was observed at Eccles on 24 May 1978 by P. Summers, and on the North Bank of Tweed below Lennel Churchyard I saw two males and two ova on *Alliaria petiolata* on 27 May 1978. The above constitute all the records for VC 81 known to me at the time of writing.

(ii) *In East Lothian VC 82.*

In East Lothian (VC 82) less seems to be known of the Orange Tip than in Berwickshire. W. Evans could only record two at Tynefield in May 1860 and 1861 (*Ann. Scot. Nat. Hist.* 1897, p. 91). It would therefore be of interest if any reader has knowledge of records this century.

(iii) *In North Northumberland VC 68.*

The decline of the Orange Tip in Berwickshire towards the end of last century was matched by a similar decline in North Northumberland (VC 68). Thus in 1839, P.J. Selby recorded it on his estate at Twizel near Belford (*Ann. of Nat. Hist.* III, 372); in 1843, Dr Johnston observed it on 3rd May on the south bank of Tweed between Horncliffe and Norham (*H.B.N.C.* II, 44) and in 1857 George Wailes noted it near Callaly on 4th June and added. "Generally distributed over the two counties", (ie. Durham and Northumberland). (*T.T.N.F.C.* III, 195.)

In 1867 the butterfly was recorded at Lilburn Tower on 28 March 1867 (a very early date), and again on 10 June 1869 at the same place by R.F. Wheeler (*T.N.H.S.* III, 28 and 478); and in 1872 it was stated to be scarce at Rothbury by R.F. Wheeler and R.E. Hooppell (*T.N.H.S.* V, 99).

George Bolam, writing in 1925, recorded it for Hetton Hall near Belford. He wrote "In W.B. Boyd's collection in 1883, I saw a considerable series all taken at Hetton Hall, where as he informed me it used to be common". (*H.B.N.C.* XXV, 522.)

In the Phenological Report of the Royal Meteorological Society for 1929 the Orange Tip was recorded for Thornton near Shoreswood Berwick and for Lemmington nr. Alnwick. It is interesting to note that it was seen at Gargunnock, near Stirling (VC 85) on the early date of 27 March 1929.

Since the above we have no further records for VC 68 until 12 June 1976, when two males were seen by P. Summers on the old railway track near Powburn. It is thus obvious that in North Northumberland, as in Berwickshire, the Orange Tip suffered a long period of eclipse.

(iv) *In South Northumberland VC 67.*

As in County Durham (VC 66), the Orange Tip apparently suffered a decline in numbers towards the end of last century but it seems never to have become completely extinct. The first record is that of John Wallis at Simonburn: "Frequent in warm vales in May and June," in 1769 (*Natural History and Antiquities of Northumberland* p. 353).

The next three known records are in a notebook of Albany Hancock in the Hancock Museum, Newcastle Upon Tyne. They are for the Newcastle area: 4 June 1826 and 29 April 1827, "Sides of lanes common": and his brother John Hancock, similarly recorded it for Ponteland Road, 3 June 1827.

George Wailes recorded it on 1 June 1860 at Riding Mill between March Burn and Dilston Castle (*T.T.N.F.C.* V,3); between 1861 and 1866, it was recorded each year at Stamfordham by J.F. Bigge and H.T. Mennell (*T.T.N.F.C.* V, 209; VI, 50); and in the same period for Burradon 26 May 1861, Cambo 19 May 1862 and Plessey Woods, 26 May 1865 (*ibid.* and *T.N.H.S.* I, 237).

Between 1867 and 1871 it was noted each year at Wallington by R. F. Wheeler, and on 16 June 1872 at Cresswell (*T.N.H.S.* V, 99). This was the last known record for the century in vc 67, but in 1899, J. E. Robson wrote, "For some years this pretty species all but disappeared but it has resumed its usual numbers" (*T.N.H.S.* XII, 4).

The earliest records I know of Orange Tips in VC 67 this century occur in a notebook of G.T. Nicholson in the Hancock Museum. For 4 June 1900 he wrote, "At Dipton near Hexham, Rosie took one male Orange Tip". Again on 28 May 1901, he wrote, "Allendale, noticed this species flying in the locality". Similarly, on 4 June 1906 at West Dipton Burn, "Orange Tips seen but not caught". It is thus certain that Orange Tips were established in South Northumberland in the first decade of the century.

George Bolam, writing in 1925, said that Abel Chapman saw several about Houxty in 1918 – "The first he had ever seen anywhere in Northumberland. Since then it has appeared about Wark, in small numbers, in most years; and almost the same may be said of several other Tynedale localities, both to east and south." (*H.B.N.C.* XXV, 522). Other early records are to be found in the Journal of W.G.

Watson for Sidwood (N. Tyne). Orange Tips are there recorded for the garden at Sidwood on 1 May 1920, 30 May 1920, 2 June 1920, 5 June 1920 and on 27 May 1920 at Redheugh Wood. In 1930, J. R. Robinson wrote "A female Orange Tip, from which I now have eggs, taken at Ponteland, with one seen by Professor Harrison at Corbridge, shows that this butterfly can still be seen in South Northumberland," (*Vasculum*, XVI, 119). In 1932, G. Bolam recorded one male in the glen below Staward Peel on 22 May 1932 *Vasculum* XIX, 123) and in 1934, F.C. Garret wrote, "Not uncommon in South Northumberland, but seems to be becoming more scarce," (*Vasculum* XX, 46). However, the interpretation may have been the exact reverse of the truth.

In the mid-thirties, Robert Craigs recorded Orange Tips in Redesdale, at Catcleugh, two on 27 May 1935 and at Rochester two on 27 May 1935 and 1 June 1935 (*H.B.N.C.* XXIX, 17). In 1941 it was observed as increasing near Stocksfield "between middle to end of June" J.W.H. Harrison (*Vasculum* XXVII, 6). In 1942 the species was again recorded for Stocksfield by Mrs T.E. Hodgkin – two seen (*Vasculum* XXVII, 32). Similarly, in 1945 it was seen at Bardon Mill on 6 May 1945 by C. J. Gent (*Vasculum*, XXX, 47), and commonly in Allendale by J. S. Ash (*Vasculum* XXX, 55). In 1946 F. W. Gardner recorded it as "Now quite common at Riding Mill," and added, "Has increased considerably of recent years," (*Vasculum*, XXXI, 6). In 1951 it was observed at Apperley Dene, 2 June 1951; in 1960, at the Sneap 21 May 1960 J. W. H. Harrison (*Vasculum* XXXVI, 11 and XLV, 11): and in 1963, F. W. Gardner again recorded it for Riding Mill as "of regular occurrence throughout the district and fairly common in favourable years," (*Vasculum*, XLVIII, 23 and *Ent. Gaz.* XIII, 22).

The increase continued in the seventies. Thus it was seen in June 1970 by J. T. B. and D. Bowman south of Caw Lough near Bonnyrigg Hall not far from Hadrian's Wall (*Vasculum*, LIX, 44); and in the same year J. D. Parrack saw one male just emerged at Plashett's Pond (N. Tyne) on 6 May 1970.

In 1971 one was seen on 1 May 1971 by A.M. Tynan at the Belling (N. Tyne). Two more were seen at Plashett's Pond by J. D. Parrack on 28 May 1971 and 6 July 1971 and in 1972, the same observer saw specimens at Plashetts on 5 June 1972, at Bolam 5 June 1972, Smalesmouth 28 April 1972 (a pupa) and at Williamstone (S. Tyne) on 28 June 1972. In 1973 several pairs were seen at Slaggyford (S. Tyne) by G. Fenwick on 9 June 1973; J. D. Parrack observed one near Creswell 11 July 1973 and three at Williamstone; in 1975 it was seen at Low Shilfurd near Stocksfield by O. L. Gilbert (*Vasculum*, LXI, 8), and also at Warden near Hexham by D. A. Sheppard and M. Eyre on 1 June 1975; and in 1976, it was noted at Stocksfield 8 June 1976 and at Close House near Newcastle on 27 May 1976 by D.A. Sheppard, and J. D. Parrack recorded six at Staward on 13 June 1976 (*Vasculum* LXI, 96).

In 1977, it was again seen at Warden on 28 May 1977 by A. Garside, D. A. Sheppard and M. Eyre, and also at Whittle Dene 4 June 1977, and was observed at Heddon on the Wall by H. T. Eales on 21 May 1977. A male was seen flying across an open bare field between Bolam Lake and Shaftoe Crags on 22 May 1977 (A.G.L.). A single ovum was found at Wallington on *Alliaria petiolata*, by P. Summers on 20 June 1977.

In 1978 the increase was much in evidence in the South Tyne and N. Tyne areas. It was seen at Brunton, Humshaugh and Haughton on 28 May 1978 by M.E. Braithwaite at Chirdon Burn, by A. M. Tynan, at the Kielder Dam site on 3 June 1978 (A.G.L.) and between Riding Mill and Slaley on 29 May 1978 by Mrs Pybus of Dipton House.

It is clear therefore that in South Northumberland (VC 67) there is evidence of almost a continued presence of the Orange Tip from the time of John Wallis (1796) up to 1978, with a possible diminution of numbers in the period between 1872 and sometime before 1899, when Robson published the first part of his "Catalogue". This partial break coincides more or less with the apparent extinction in Berwickshire lasting the greater part of a century.

(v) *In County Durham (VC 66)*

As early as 1846, the Orange Tip was recorded by Rev. George Ornsby in his book *Sketches of Durham*. Later George Wailes (1857) described it as "generally distributed" (*T.T.N.F.C.* III, 195). The evidence for the rest of the last century is inconclusive; in all probability the species persisted though it may have decreased towards the end, as in Northumberland and the Scottish Border counties.

During the first two decades of the present century the Orange Tip seems to have become scarce. In 1939, Professor J.W. Heslop Harrison published an account "The present position of our local butterflies compared with that of 40 years ago" in which he wrote: "In 1900 this butterfly occurred in certain Durham and southern Northumberland localities in small numbers, and even the years 1902 and 1903 produced little change for the worse. Nevertheless, after that its numbers progressively lessened, and it even vanished from some localities," (*Vasculum*, LII, 119). In 1951 the same author wrote, "After many years of decadence produced by the bad seasons of 1902–3 the Orange Tip began to recover lost ground in 1919. One of the last areas to be recolonized was the Blackhall Rocks. However, on July 2nd it was flying in the little dene half a mile south of the Hotel", (*Vasculum* XXXVI, 24). As early as 1905 G.T. Nicholson had recorded in his field notebook that he netted nine specimens in June at Croft.

Its history in the years 1919–1969 is one of regular occurrence, generally at low density but in relative abundance in some seasons in certain localities. Writers regarded its years of plenty as welcome exceptions to its general scarcity. During this period, the date of its first appearance was recorded under the Royal Meteorological Society's scheme for phenological observations. Such a record testifying to its presence was made in fourteen of the fifteen years 1925–1939 by the Darlington and Teesdale Naturalists' Field Club (*Vasculum* XXVI, 20–24). Years of plenty include 1925, 1930, 1934, 1935, 1937, 1939 and 1945, only seven years in the period under consideration. Throughout the fifties and early sixties no marked change seems to have been noted. Thus in 1958, Professor J. W. H. Harrison wrote, "This butterfly was very scarce this year although a female was found at rest on flowers of an umbellifer at Elemore on June 28th" (*Vasculum* XLIII, 32). The first indication of marked increase was noted in 1969 by T. C. Dunn who wrote, "The Orange Tip was much more widespread than usual" (*Vasculum* LIV, 10). Evidence of the increase in the Redworth area in 1971 was given by T.W. Jefferson who wrote, "Some species have appeared for the first time in this area including *Anthocharis cardamines* in places where they have been strangers since 1965" (*Vasculum* LVII, 1). In 1973 T. C. Dunn wrote "This year has seen a sudden increase in the population of the Orange Tip butterflies in Durham County . . . all inland colonies seem to have increased, with some of them spilling over their natural boundaries . . . (*Vasculum* LVII,1). This year it has turned up all over the place and has even appeared in urban gardens in some of the most built-up industrial towns. Has the recent series of mild winters had anything to do with it?" (*Vasculum* LVIII, 10.) In a later note he added, "The first specimen noted in the Chester-le-Street area was in my garden on May 25th 1973. This is right in the built-up area of the town and this is the first time I have known it to fly so far from its usual haunts in 25 years residence here . . ." (*Vasculum* LVIII, 13.) Similar reports were recorded from Darlington, south-west Durham, Winlaton Mill and elsewhere so that T.C. Dunn again wrote in 1975, "News has been flowing in of its increase almost everywhere in Durham . . ." (*Vasculum*

LX, 4.) Subsequent records show that this increase and spread has continued into the present year (1978).

(vi) *In Peebles (VC 78); Selkirk (VC 79) and Roxburgh (VC 80).*

In order to try and trace the increase of the Orange Tip in these Scottish border counties, I have sought the help of Mr Arthur J. Smith (A.J.S.) of Selkirk and Mr Andrew G. Buckham (A.G.B.) of Galashiels, from whom I have received most of the records listed below.

At Peebles one female and two males were seen on 18 July 1975 and several at Innerleithen on 18 May 1978 (A.G.B.). Doubtless there would be earlier records and it would be of interest to know of these, especially if they are for last century.

For Selkirkshire, A.J.S. wrote saying he knew of no records before 1975. His former schoolmaster who was a keen collector had never seen it. Mr Smith's first view of a Border specimen was one which came to him in a matchbox from Lanton in Roxburgh about 1928. His first known record for Selkirk was on 17 May 1975 at Duchess Drive, Bowhill, and the second of a male seen by Dr Meikle at Bridgeheugh on 27 May 1975. Another male was seen in a Selkirk garden by Dr C. Tinlin on 30 May 1975. Ova were found on *Alliaria petiolata* at Bridgeheugh near Selkirk on 3 June 1976 (A.J.S.) and two males were seen at Thirlestane, Ettrick, in 1976 by Mrs Sanderson. Several were seen at Mauldsheugh by the river at Selkirk on 26 May 1977 (A.J.S.). One male was seen at St. Mary's Loch on 25 May 1978 and another at Selkirk Hill on 27 May 1978 (A.J.S.). Several males were seen at Blackpool Moss on 3 June 1978 and at Nether Whitlaw Moss on 4 June 1978 (A.G.B.). Others were observed at Lindean Reservoir 4 June 1978 (A.J.S.).

For Roxburghshire the published evidence shows that the Orange Tip was well established in the second half of last century. In 1867 Sir Walter Elliot recorded it for Denholm (*H.B.N.C.* V, 329), and in 1882 Adam Elliot of Samieston, Jedburgh recorded it for Roxburghshire (*H.B.N.C.* X, 154). In 1895 William Grant Guthrie reported it as not uncommon for Burnfoot and Hornshole at the east end of Old Hawick (*H.B.N.C.* XV, 332). George Bolam in 1925 wrote that "More than fifty years ago it used to be taken fairly commonly by W.B. Boyd at Cherrytrees near Yetholm." (*H.B.N.C.* XXV, 552). By 1901, it was once more becoming frequent in Jedwater (*Entomologist* LXIII, 130). Mr A.J. Smith received a specimen from Lanton about 1928 (see above).

More recently there has been a remarkable increase of the Orange Tip in Roxburgh similar to that in Northumberland and Durham. In the years 1971–1978 it has been seen each year at Wells Sawmill near Denholm (A.G.B.). Other records are from Crookholm Wood in Liddesdale in June 1977 (A. Garside). In 1978 it was seen at St. Boswells (P. Summers and A.G.B.), Newtown St. Boswells and Maxton (A.G.B.), Denholm (G.A. Elliot), Morebattle (M.E. Braithwaite) and Kelso and Galashiels (A.G.B.), between May 17 and June 3.

It would appear therefore that the commencement of the increase in the Scottish border counties was about 1971, two years later than in County Durham, but four years earlier than that in Berwickshire. This might mean that the increase in the Borders was in part caused by a spread from the south or west in a north and easterly direction.

(vii) *In south-west Scotland (VCs 72–74)*.

Although I have no personal knowledge of the Orange Tip in south-west Scotland, the late Mr E.C. Pelham-Clinton of the Royal Scottish Museum, has informed me that there are a few dated records e.g. for Stranraer (VC 74) in 1882; Kirkcudbrightshire (VC 73), in 1884; followed by a gap up to 1943 when it was recorded for Tynron, Dumfries (VC 72), and several records in 1947 and 1948. This pattern suggests a decline and recovery.

(viii) *In Cumbria VCs 69 and 70*.

According to information received from Dr N.L. Birkett, the Orange Tip has been common in Southern Cumbria over a period of about fifty years. This is confirmed from notebooks of the late Dr R.C. Lowther of Grange-over-Sands who also maintained that it was scarcer in northern Cumbria (formerly Cumberland). The late Frank Littlewood of Kendal also recorded it in the Kendal area (common) but said it was scarcer about Lancaster. The evidence therefore suggests that in at least southern Cumbria as in Durham and south Northumberland, the Orange Tip has been present for as long as records are known. In northern Cumbria, however, its status is less certain. G.T. Nicholson recorded Orange Tips near Keswick (VC 70) on 4 June 1911, and in a wood near Gilsland railway station near the north-eastern limit of Cumbria on 25 May 1896. I personally saw a few specimens near the King Water in VC 70 on 28 May 1977.

Conclusion

The evidence shows that in the Scottish Border counties the Orange Tip was established last century, but a decline set in sometime after 1880. This resulted in virtual extinction over much of the Border region but mostly in the eastern half. Except in certain years, this state of affairs lasted up to the seventies of this century. The increase that then occurred, was possibly the culmination of a long gradual recovery most pronounced in the south and west resulting in a steady extension northwards and eastwards. Two factors may have influenced this recovery.

Mr E.C. Pelham-Clinton favoured the view that agriculture, especially draining and grazing may have gradually eliminated *A. cardamines*, but that forestry has possibly helped it to spread again, since *Cardamine pratense* grows along the sheltered drainage ditches in young plantations. Another factor may be the more recent changed policy of reduced cutting of roadside verges. This favours the growth of *Alliaria petiolata* another favourite food plant of the Orange Tip along hedgerows and roadside ditches.

Whether or not the climate was a factor affecting the decline of *cardamines* towards the end of last century is impossible to assess. There were two very severe winters in the Scottish Borders in 1879 and 1880, but the effect of these on the Orange Tip population was apparently never noticed to my knowledge.

Abbreviations

> H.B.N.C. History of the Berwickshire Naturalists' Club
> T.N.H.S. Transactions of the Natural History Society of Northumberland, Durham and Newcastle-upon-Tyne.
> T.T.N.F.C. Transactions of the Tyneside Naturalists' Field Club.

Since writing the above my attention has been drawn to the fact that within the last decade the average mean temperature for the month of April has fallen by about one Fahrenheit degree. This has probably pushed the time of emergence of the Orange Tip imagines from the chrysalis, into the month of May when the weather is usually more suitable for the mating and egg laying of the Orange Tip. This may have been a factor leading to the return of the Orange Tip butterfly.

CHAPTER 5

Articles on General Botany

(a) Presidential Address 1972.
(b) On the Re-discovery of *Linnaea borealis* at Mellerstain.
(c) The 1974 Botanical Meeting.
(d) Natural History Observations in 1979.

(a) **Presidential Address.** *Delivered to the Berwickshire Naturalists' Club at Berwick, 4th October 1972. From the History of the Berwickshire Naturalists' Club, Vol. XXXIX. Part 2, 1972, pp. 85–96.*

The Early History of Seeds

I. *Introduction. What is a seed?*

It is customary in all scientific subjects to define the terms and units used, so by way of introduction I think we should first consider what a seed is. The literal meaning of the term seed is 'that which is sown'. In botany, however, the term is used only for the unit of reproduction in higher plants. In lower plants the unit of reproduction is typically a spore. We therefore distinguish between spore plants and seed plants.

A spore is typically a single specially resistant cell which is set free and dispersed and from which a new plant can grow if given the necessary conditions. In contrast a seed is a multicellular body.

A spore is usually formed asexually i.e. without the union of two special sex cells. In contrast a seed is the product of both an asexual and sexual process involving the formation of a fertilized egg and food store.

Examples of spore plants are the ferns, horsetails, club-mosses, true mosses, liverworts, fungi and some algae and bacteria. Examples of seed-plants are the Gymnosperms and Angiosperms. Gymnosperms have naked seeds i.e. the seeds are not enclosed in an ovary or fruit. Our three native species of Gymnosperms are the Scots Pine, Yew and Juniper.

Angiosperms have enclosed seeds i.e. they develop inside a gynoecium built up of one or more carpels.

Let us consider what happens in the life cycle of a fern as an example of a spore plant. The spores are usually produced in sporangia borne somewhere on the fronds e.g. on the lower surface. Such a plant producing spores is called a sporophyte. When a spore germinates it does not grow directly into another fern sporophyte but forms a small green plant called a prothallus or gametophyte. It is called a gametophyte because it produces gametes or sex cells. After the fertilization of an egg by a sperm an embryo forms and this commences growth, at first drawing food from the prothallus. Eventually it becomes independent and has roots, stem and leaves. Thus in the life-cycle of a fern there are two generations and typically these alternate, the sporophyte reproduces asexually by spores, and the gametophyte reproduces sexually by gametes. In this way the fern gets the best of both worlds; it has the advantages of reproduction by spores eg. wide dispersal, and the advantages of sexual reproduction enabling variation and adaptation to occur.

In plants which live on land there is always one great disadvantage involved in sexual reproduction: this is the need for liquid water to make fertilization possible. The male gamete needs water in which to swim in order to reach the female gamete. It is this disadvantage the seed habit overcomes. The primitive seeds possessed a pollen chamber containing a watery liquid which acted as a private pond ensuring that the male gametes had the necessary liquid in which to swim and bring about fertilization.

In most ferns the spores of any one species are all alike in average size and form. Such ferns are said to be homosporous. In some other spore-plants as for example the Lesser Clubmoss *Selaginella spinulosa* (which grows on Greenlaw Moor and elsewhere), the spores are of two kinds – larger megaspores and smaller microspores. From these develop separate male and female gametophytes. Megaspores produce female gametophytes and microspores produce male gametophytes. In this way more food is provided in the prothallus which may bear a future embryo sporophyte. Spore plants which produce the two kinds of spores – megaspores and microspores – are said to be heterosporous.

We are now in a better position to understand what a seed is. A seed is a megasporangium producing only one functional megaspore. This megaspore, however, is never released, but germinates to produce an internal, reduced, female gametophyte on which female organs known as archegonia form egg cells from which, after fertilization, embryo sporophytes can be formed. In addition to the megasporangium a seed has an integument, a kind of protective coat which is called the testa. This integument is two layered in most angiosperm seeds, but single-layered in gymnosperm seeds.

In modern seeds it is customary to call the megasporangium the nucellus, and the megaspore is called the embryo sac. In the seeds of Gymnosperms such as Scots Pine the female prothallus acts as a food reserve called endosperm but in Angiosperm seeds the prothallus is reduced beyond recognition and the food reserve or endosperm only develops after fertilization.

By the time a modern Angiosperm seed is set free from its parent plant it consists of parts of three generations telescoped into one reproductive structure. These are

1. The megasporangium and its integument representing the first sporophyte generation.
2. The prothallus or female gametophyte generation, with the egg cells representing the next gametophyte generation.
3. The embryo or baby plant developed from a fertilized egg which represents the next sporophyte generation.

This analysis enables us to define the terms 'ovule' and 'seed' as follows: An ovule is an integumented megasporangium retaining a single functional megaspore.

A seed is a matured ovule which may or may not contain an embryo plant at the time of shedding.

From this we see that seed-plants appear to have evolved from spore plants and the chief interest of fossil seeds is the light they shed on how this evolution took place.

Primitive seeds are more like simple megasporangia than modern seeds. Those of the Pteridosperms or seed-ferns very often show no sign of a female prothallus inside. When the prothallus is present within the megaspore the seed coat shows signs of wear and tear as if the seed had been dropped some time before the female gametophyte had developed. So far no embryo has been found within a Pteridosperm seed though pollen is often present inside the pollen chamber. That which is sown is therefore a mature ovule which has usually been pollinated. Development of the female gametophyte and fertilization probably occurred while the seed was lying on the ground after abscission. Once the embryo started forming from the fertilized egg there was probably no further dormant period. It had to go on developing.

The method of pollination in primitive seeds was probably by wind, the pollen being caught in a droplet of liquid exuded at the apex of the ovule followed by resorption into the pollen chamber. Afterwards the pollen grains must have set free the male gametes capable of swimming in the fluid of the pollen chamber to fertilize the eggs on the female prothallus as happens in the modern *Ginkgo biloba* or Maidenhair Tree which on this account can be called a living fossil.

II. *What plants could have given rise to the seed habit?*

The oldest known fossil seed is called *Archaeosperma arnoldii*. It was found relatively recently in the Upper Old Red Sandstone of Scaumenac Bay in the Province of Quebec, Canada (Pettitt and Beck 1968). It is preserved as a compression so that its internal structure is not known. The seeds occur in groups of four within a cupule and are associated with compressed fronds of *Archaeopteris*. The oldest known petrified seeds at the time of writing, showing the internal structure preserved come from the Lower Carboniferous rocks of the Cementstone Group of Berwickshire and East Lothian. Most of these seeds were probably borne on Pteridosperms, though for most the parent plant is not known. Plants thought to be Pteridosperms occur in the Upper Devonian rocks so that most palaeobotanists think that they must have evolved in the Devonian period.

Pteridosperms have fern-like foliage so that they are sometimes referred to as the seed-ferns. It is very unlikely, however, that they evolved from ferns. Ferns do not typically have secondary wood as do the Pteridosperms.

Fossil plants with fern-like leaves, secondary wood and reproduction by spores are now classified as Progymnosperms or Progymnospermopsida. *Archaeopteris* with its large fern-like fronds was once thought to be a fern but is now classified as a progymnosperm and was a huge tree. The petrified trunks were known for a long time under the name *Callixylon* before the discovery that they bore the large fronds of *Archaeopteris* (Beck 1960). It now seems most probable that the seed habit evolved in the Devonian Period among Progymnosperms such as *Archaeopteris* or its relatives and not in the ferns.

In the British Isles the best locality for *Archaeopteris* is at Kiltorkan in Southern Ireland. In the Upper Old Red Sandstone of Scotland *Archaeopteris* is very rare but specimens now in the Royal Scottish Museum were obtained last century by a Mr John Stewart and given to Hugh Miller. They were said to come from a quarry at Preston Haugh near Duns in Berwickshire. The site of this quarry I have never discovered and I have sometimes wondered whether or not the specimens were actually obtained from the Old Red Sandstone scaur on the right bank of the Whiteadder between Preston Haugh and Barramill Plantation now largely felled. This scaur is on the right bank of the river on the first big bend below Cockburn Bridge, and here I have found compressions of large stems which although without leaves could well be *Archaeopteris*. One of Mr Stewart's specimens was figured by Miller in his book *The Testimony of the Rocks* (Miller 1869, p. 411).

Before Pettitt and Beck described the compressed seeds of *Archaeosperma arnoldii* from Canada some isolated megaspores were described under the name *Cystosporites devonicus* from the same locality (Chaloner and Pettitt 1964). These were identical with the megaspores in the seeds of *Archaeosperma* and like them were associated with fronds of *Archaeopteris*. It is of interest that one megaspore agreeing with *Cystosporites devonicus* has been found in Berwickshire (Long 1968). It was found in stratified volcanic ash in the bed of the Whiteadder above Preston Bridge near Cumledge and not very far from the site where Mr Stewart's specimens described by Hugh Miller were probably found.

So far no one has proved that any species of *Archaeopteris* bore seeds but it is known that some species of *Archaeopteris* were heterosporous having some sporangia producing megaspores and others producing microspores.

(III) *How could a seed have evolved from a sporangium?*

The changes involved must have been fourfold:

1. The establishment of heterospory and reduction in the number of functional megaspores to one in each megasporangium.
2. The incomplete dehiscence of the megasporangia so that the megaspores were never released.
3. The development of a pollen chamber inside the top part of each megasporangium and capture of pollen in a droplet of exuded liquid.
4. The development of an integument from sterile branchlets borne near the megasporangium

The simplest hypothetical starting point for the evolution of a seed is the terminal sporangium in the Psilophytales such as *Rhynia*. These primitive vascular plants were homosporous.

The evolution of heterospory is exemplified by *Archaeopteris*, eg. in *A. latifolia* (Arnold 1939).

Reduction in numbers in the megaspores in a single megasporangium is exemplified by *Stauropteris*. This is a primitive fern-like plant of which three species are known. In the lower Carboniferous species each megasporangium was probably shed intact with two megaspores and two smaller spores inside. The sporangium had a terminal beak which may or may not have been opened (Surange 1952).

The evolution of an integument is best seen in *Genomosperma* (Long 1960) where eight sterile cylindrical branchlets had commenced to fuse around the base of the megasporangium. Later a micropyle evolved by the integumental lobes fusing completely around the top of the megasporangium.

(IV) *How could the second integument of Angiosperm ovules have evolved?*

Most Angiosperm ovules and seeds have a double integument whereas in Gymnosperms the ovules have a single integument. One theory put forward to account for the second integument is to regard it as derived from a cupule. Certain Pteridosperms did possess single-seeded cupules and for these the cupular origin of the second integument seems feasible, at least for ovules that are orthotropous. Most Angiosperm ovules, however, are curved. When the ovule body is curved roughly at right angles to the stalk or funicle the ovule is said to be campylotropous, as for example in the Wallflower. If the ovule body is reflexed so that it is inverted and fused on one side to the stalk it is said to be anatropous as in a Buttercup. If such anatropous ovules are sectioned longitudinally, it is seen that the second integument is present only on the surface away from the stalk. This seems to suggest that the second integument was not present before the curvature took place or it would be represented all round the megasporangium like the first integument. It is therefore more likely that the second integument formed either during or after the time when the curvature took place. At this time of writing my view is that the second integument was not derived from a cupule but more probably resulted from a differentiation within the first integument (NB later this view had to be abandoned in favour of the view that the second integument was cupular in origin). I think that curvature of primitive seeds occurred mainly in those having only two integumental lobes. During curvature the two free lobes could have become differentiated as an inner integument which was overhung by a dorsal lip. In this way the pollination droplet would be better protected from rain by the hood-like dorsal lip or so-called second integument.

V. *How did seeds become enclosed in carpels as in modern flowers?*

In modern Angiosperms the female part of the flower is known as the gynoecium and it consists of one or more carpels. As an example let us consider the flower of a Sweet Pea *Lathyrus odorata* L. Here the gynoecium is monocarpellary i.e. it consists of a single carpel. The carpel has three parts known as the ovary, style and stigma. Inside the ovary are the ovules. After pollination the ovary ripens to form a fruit and the ovules become seeds. The style and stigma wither away. The fruit of the Sweet Pea is a pod and this has dorsal and ventral sutures which split open eventually to release the seeds.

In a flower such as the Marsh Marigold *Caltha palustris* L. the gynoecium consists of several free carpels. These ripen to form a fruit known as an aggregate of follicles. Each follicle is very like a pod having dorsal and ventral sutures but eventually only one of these opens to release the seeds. This is the ventral suture. A follicle like that of the Marsh Marigold is thought to be one of the more primitive forms of carpel. Such a carpel has a very short style and stigma on top of the ovary but in some Angiosperms thought to be even more primitive the stigma occupies the entire length of the ventral suture and really consists of the two adpressed margins of the carpel.

The term "suture" really means a "seam" and the ventral suture really consists of the two margins of the carpel pressed together and fused. There is good evidence that at one time there was a ventral opening down this side of the carpel and in some living Angiosperms this can still be seen. Thus in the Dyer's Rocket *Reseda luteola* L. which grows fairly commonly on shingle by the Whiteadder and elsewhere, the ventral suture is open near the top so that we can see the ovules inside the carpels of the flower. This gymnospermous character supports the idea that Angiosperms have evolved from Gymnosperms. Originally each carpel must have been like a purse with a slit down one side through which pollen could be blown by the wind so gaining direct access to the ovules for fertilization. The closure of the ventral suture probably began basally like a zip fastener formed of interlocking hairs. The prime function of these hairs was to reduce loss of water by evaporation from the delicate ovules. Such hairs however would impede the entry of wind-borne pollen-grains which would become entangled among the hairs. At some time such pollen grains acquired the power to germinate and form a pollen tube through which moisture was absorbed. Eventually the pollen tube became the structure through which the male gametes were passed to the egg-cell without being released. This mode of fertilization is known as siphonogamy.

VI. *From what organ was the carpel derived?*

Among Pteridosperms the seeds, although naked, were often protected by an enveloping structure known as a cupule. In some, each seed had its own individual cupule and several cupules were borne on a single leaf or frond. In others, two or more seeds occurred within one cupule and again several of these were borne on a single frond.

In a third type an entire frond was converted into a single relatively large cupule about the size of a Tulip flower-bud. One such example is called *Calathospermum* and the species known as *Calathospermum fimbriatum* (Barnard 1960) was first described from rocks at Oxroad Bay near Tantallon Castle on the East Lothian coast. It has also been found in limestone nodules occurring in the bed of the Whiteadder between Edrom and West Blanerne. In most Pteridosperms the leaf was a large frond which forked like a letter Y. This forking of the leaf-stalk or petiole is represented in the base of a *Calathospermum* cupule so that it consists of two lateral halves borne on a common stalk. Now such a bivalved structure may be compared to the two halves of a single carpel such as a pea-pod. It is therefore possible that the carpel in modern flowers really represents a highly modified cupule in which the two halves have joined so as to enclose and protect the seeds up to the time of their release.

For this and other reasons some palaeobotanists think that Pteridosperms were the most likely ancestors of the Angiosperms.

A modern flower is really a telescoped shoot in which there are basically three sorts of lateral organs borne on a stem-like axis called the receptacle. The three sorts of lateral organs are the carpels, stamens and perianth leaves sometimes differentiated into sepals and petals. Goethe put forward the idea that all these organs are really modified leaves. It is possible, however, that the carpels and stamens evolved from branched cylindrical organs bearing sporangia which have never become typical flattened leaves.

Goethe's idea was that a carpel is really a leaf folded down its midrib so as to enclose the ovules borne along its margins. If on the other view, the ovules represent megasporangia borne on cylindrical stalks it seems more likely that these fused basally, whereas in the evolution of the true leaves the fusion of simpler units seems to have been by webbing between terminal parts, so giving flattened portions which were better adapted to absorbing light for photosynthesis.

By the study of fossil seeds and the plants that bore them the mystery of how flowering plants evolved may yet be solved. Fossil Angiosperms go back in the rocks only to the early Cretaceous period about 135 million years ago. The Lower Carboniferous seeds found in Berwickshire and East Lothian are almost 350 million years old. Between the end of the Carboniferous Period and the commencement of the Cretaceous Period was a span of 135 million years covering the Permian, Triassic and Jurassic geological periods.

The answer to the riddle of the Angiosperms must lie somewhere locked up in these rocks unless their ancestors completely evaded any natural burial and like Enoch of old passed from this life without leaving any physical remains behind (Genesis 6:24).

(b) **On the re-discovery of Linnaea borealis at Mellerstain.** From the *History of the Berwickshire Naturalists' Club, Vol. XXXIX part 2 1972, pp. 142–3.*

Linnaea borealis was first discovered at Mellerstain by Mr Dunn the gardener in 1834 (*H.B.N.C.* I, 248). It was first seen there by members of the Club in 1843 and Dr Johnston wrote that it occupied two or three considerable patches located in a fir wood on Lightfield Farm (*Natural History of the Eastern Borders*, 1853, p. 99). In 1866 the plant was seen again at the same site on 28th June, and said to be in full flower (*H.B.N.C.* V, 244).

In 1869 Dr Charles Stuart described the plant as growing over an area which was about eighty paces in circumference on 2nd July (*H.B.N.C.* VI, 71).

In 1880 James Hardy noted that the plant was still present at Mellerstain (*H.B.N.C.* IX, 229 and 293).

In 1894 Rev. G. Gunn recorded its presence at this locality (*H.B.N.C.* XV, 82). In 1915, Rev. J.J.M.L. Aiken could not find it at Mellerstain (*H.B.N.C.* XXII, 354), but in the following year it was found by J. Ferguson (*H.B.N.C.* XXIII, 47). This is the last record I have found for the plant at the Mellerstain site.

On 1st July, 1972, the Club held a special botanical meeting at Mellerstain with the object of trying to relocate the plant. Mrs E.O. Pate made the arrangements, first obtaining permission for the visit from the estate factor, Mr John E. Hume. Mr Sturrock, the Forester at Mellerstain, kindly met the party of about twenty-five members who assembled at the "Cocked Hat" plantation at the eastern entry to the estate on the road south from Gordon. As Bonaparte's Plantation had been felled and re-planted in relatively recent years, fears were entertained that the *Linnaea* might have died out, since it is a shade-loving plant. Much time was spent without success searching among the herbage between the young trees of the new plantation but a few members ventured over the wire-netting fence at the west side and near to a ditch bounding the older birch wood which had not been re-planted. It was in the shade of this wood, a short distance from the NE corner and due south of Lightfield Farm, that Messrs. J. and G. Waldie of Gordon found the plant growing over an area of a few square yards and bearing its twin bell flowers of delicate pink colour.

The plant was photographed and the whole party summoned to see the plant which had survived over the span of 138 years from its initial discovery. Nearby a Woodcock was flushed from its nest with four eggs. Everyone went away gratified to know that this interesting plant named in honour of the great Swedish botanist was still surviving in the place of its first discovery in Berwickshire.

Other sites at which *Linnaea* has been found in the County are Huntly Wood near Gordon (1880, *H.B.N.C.* IX, 294); Longformacus strip (1884 *H.B.N.C.* X, 608); wood between Drakemire and Brockholes (1891, *H.B.N.C.* XIII, 386) and Fans (1922, *H.B.N.C.* XXIV, 358). It would be of interest to investigate these localities and determine whether or not the plant still survives at these stations also.

(c) **The 1974 Botanical Meeting. From the** *History of the Berwickshire Naturalists' Club, Vol. XL p. 58.*

A botanical meeting was held on Saturday, 17th August 1974 at Cattleshiel and Wellcleugh Burn, to look for *Saxifraga hirculus*, the Yellow Marsh Saxifrage. The President Mr Adam R. Little, together with about twenty-six members and visitors met at 10.30 a.m. on the Westruther road between Foul Burn Bridge and Cattleshiel Farm. Prior arrangements had been made by Mrs E.O. Pate, who obtained the necessary permission from the three farmers on whose land the search was made. The morning was bright and sunny with a westerly breeze, and although one or two showers threatened rain this was very slight and in no way spoiled the proceedings.

Members were first shown herbarium specimens of the rare saxifrage collected last century by Robert Castles Embleton, surgeon, and R.B. Bowman together with Dr. Johnston, from the Langton-lees site. R.C. Embleton, surgeon, was a founder member of the Club and his specimens are of special interest since they are dated 1832 – the year in which the plant was first found by Mr T. Brown whose father was Minister of Langton Parish. The data label on this sheet reads "*Saxifraga hirculus*, Langton Lees Farm, Berwick, Mr T. Brown and R. Embleton 1832". In addition two sheets of *Sedum villosum* Hairy Stonecrop and *Sagina nodosa* Knotted Pearlwort were also shown, as these plants frequently occur in the same kind of habitat as the Saxifrage. A rapid resumé of the history of the plant in Berwickshire was then given.

In 1832 it was discovered by Mr T. Brown at the Langton Lees Cleugh site, (*H.B.N.C.* I, 9, 29, and New Statistical Account, 1845, p. 236).

In 1853 Dr Johnston wrote in his book *Natural History of the Eastern*

W. Ryle Elliot (left), former Secretary of the Berwickshire Naturalists' Club, with his sister Grace A. Elliot M.B.E. (centre) and David G. Long (right) on the left bank of the Tweed below Birgham opposite Carham on the English side.

Borders, p. 84: "In a wet moorish spot near Langton Wood, plentiful – Rev. T. Brown who had the good fortune to add this beautiful species to the Flora of Scotland."

In 1867 it was again recorded by Dr Frances Douglas "In a wet spongy bog, Langton Lees Cleugh. Found by Dr. Clay" (*H.B.N.C.* V, 300).

In 1872 Dr C. Stuart wrote: "In great beauty, on 14th August on the sides of sheep drains on left hand side going up Langtonlees Dean," (*H.B.N.C.* VI, 437, also in 1873, H.B.N.C., VII, 19).

In 1873 Dr C. Stuart wrote this warning. "It is a duty I owe to the Club to inform the members that *Saxifraga hirculus* is all but extinct at Langtonlees Dean in consequence of a number of sheep drains having been constructed through the place where it used to grow. My second son and Captain Norman were up at the station last summer and came back with hardly a vestige of a specimen. It, however, may spring again, but I am not sanguine." At one period it was plentiful (*H.B.N.C.* VIII, 533). In spite of this the plant still continued to grow, thus in 1879 Mr C. Watson recorded it again at Langton Lees Cleugh, (*H.B.N.C.* IX 49), and in 1883 it was found at Langton again, by Mr Arthur Evans, (*H.B.N.C.* X, 264). This was the last recorded occurrence of the plant at the Langton site although in 1886 Dr C. Stuart wrote in the *Transactions of The Edinburgh Botanical Society* Vol. 6,

19–26; "Rev. T. Brown found *S. hirculus* – new to the Scottish flora in a bog where sheep drains threaten to destroy it. It grows with a profusion of *Sedum villosum* and *Spergula nodosa*."

In 1885 James Hardy recorded the plant at a second site only about two miles distant from the first site. Thus he wrote: "Mr Robert Renton has picked up on the Cattleshiels Moor the rare *Saxifraga hirculus*. There would be over a hundred plants at the place"

(*H.B.N.C.* XI, 68). This is the last record of the species in the County. Thus in 1916 A.H.Evans wrote "Extinct at Langtonlees, in the locality originally recorded" (*H.B.N.C.* XXIII, 223). Again in 1931 Alan A. Falconer wrote: "Apparently this plant has been extinct for many years in the station in the Wellcleugh, Langtonlees, where it was discovered nearly a hundred years ago by Mr (afterwards Rev) T. Brown. I have frequently visited the spot at its flowering season in vain."

The party first followed the path which crosses the moor to Dronshiel and then diverged to the left and spread out over Shiningpool Moss to the Under Bog approaching the Big Dirrington. The expanse of likely territory here was too large to cover adequately and another hour could easily have been spent. The land has been drained by sheep drains and although these have a nice flora no sites with *Sedum villosum* or *Sagina nodosum* were found. Flowering plants encountered were: *Ranunculus acris* Meadow Buttercup; *R. flammula* Lesser Spearwort; *Drosera rotundifolia* Round-leaved Sundew; *Veronica scutellata* Marsh Speedwell; *V. officinalis* Common Speedwell; *Linum catharticum* Purging Flax; *Mentha aquatica* Water Mint; *Pedicularis sylvatica* Lousewort; *Erica tetralix* Cross-leaved Heath; *Prunella vulgaris* Self Heal; *Potamogeton polygonifolius* Bog Pondweed; *Juncus squarrosus* Heath Rush; *J. articulatus* Jointed Rush; *J. bulbosus* Bulbous Rush; *Carex nigra* Common Sedge; *C. echinata*, Star Sedge; *C. lepidocarpa* Long-stalked Yellow Sedge; *C. demissa* Common Yellow Sedge; *Scirpus caespitosus* Deer Sedge; *Sieglingia decumbens* Heath Grass; *Molinia caerulea* Purple Moor Grass.

Among insects seen were *C. graminis* the Antler Moth of which the males were very active flying in the sunshine; *C. pamphilus* the Small Heath butterfly – abundant; *A. urticae* the Small Tortoiseshell; *M. jurtina* the Meadow Brown – common; *C. didymata* the Twin-spot Carpet; and the larvae of *L. quercus* var. *callunae* the Northern Eggar; *M. rubi* the Fox Moth, and *S. pavonia* the Emperor Moth. A single specimen of the coppery green beetle *Carabus nitens* was caught. A lizard was seen and the sloughed skin of an adder. A solitary Golden Plover was heard making its mournful call-note.

After a picnic lunch taken sitting in the hot sunshine and sheltered by a dry stone dyke a move was made to Camp Moor and the Wellcleugh Burn. Plants found here were Iris pseudacorus the Yellow Flag – in large patches; *Myosotis caespitosa* Tufted Forget-me-not, all along the Burn; *Lysimachia nemorum* Yellow Pimpernel, *Pimpinella saxifraga* Burnet Saxifrage; *Salix aurita* the Eared Sallow; *Empetrum nigrum* Crowberry; *Populus tremula* Aspen; *Briza media* Quaking Grass. In the Lees Cleugh were found *Phillitis scolopendrium* Hart's Tongue Fern; and *Campanula latifolia* Giant Bell flower. A female *P. icarus* the Common Blue butterfly was seen at rest and a few *P. napi* the Green Veined White were on the wing along with many Meadow Browns.

(d) **Natural History Observations during 1979.**

On 10 March 1979, four Short-eared Owls were seen hunting by daylight about 5 p.m. along a 1 mile stretch of the A697 near Lilburn Glebe not far from Roddam between Wooler and Powburn (NU 034223). Doubtless they were winter immigrants preparing for return to Scandinavia.

On 23 March, a large squid (*Loligo forbesi*) eighteen inches long was washed up at Seaton Sluice.

The long hard winter extended into April and delayed the arrival of migrant birds. I first saw Swallows at Powburn on 25th April but on 1st May we had more snow showers. Sandpipers and Willow Warblers were back near Wooler Water at Haugh Head on 6th May and the Green Veined White butterfly was seen at Coldstream on the same day.

The Muslin Ermine *Diaphora mendica* was taken by Mr G. Evans, warden at St. Abbs Head in July. It was first taken in Berwickshire at Birgham House by Miss G.A. Elliot in 1960 and again in 1961 and 1962 (see *H.B.N.C.* XXXVIII, p. 136). It is widespread but never very common in Northumberland as evidenced by records from the file at the Hancock Museum – Jesmond NZ 26, 1899 (Mr Henderson per J.E. Robson); Kenton NZ 26, 1899 (Mr Rhagg per J.E. Robson); Sidwood NY78, 1922 and 1938 (Mr Clegg per R. Craigs): Shoreswood NT 94, 1899 (G. Bolam): Bolam NZ 08, 1925 (G. Bolam); Riding Mill NZ 06, 1962 (F.W. Gardner); Longwitton NZ 08, n.d. (R. H. Benson); Benwell NZ 26, 1969 (H.T. Eales); Prestwick Whins NZ 17, 1913 (D.A. Sheppard and M. Eyre); Throckley Pond NZ 16, 1977 (D. A. Sheppard and M. Eyre); Whitley Bay NZ 37, 1978 (J. D. Parrack); Gosforth, NZ 27, 1979, (M.A. Walker); Bedlington NZ 28, 1979 (P.R. Paul).

Six larvae of the Bedstraw Hawk Moth *Hyles gallii* were seen on

Holy Island on 26 August 1979 and 2 September 1979 by P.W. West and D.P. Hammersley. They were feeding on Rose-Bay Willow-herb *Chamaenerion angustifolium*. Fifteen larvae of the Elephant Hawk moth were also seen on the same food-plant on the island.

On 11 August 1979 a patch of Rosebay Willow-herb *Chamaenerion angustifolium* var. *brachycarpum* was found growing at Oxroad Bay E. Lothian on the steep bank below the path a little above high tide level. The only other site I have seen for this variety is the right bank of the Whiteadder about a quarter mile below Blanerne Bridge. In this variety the fruits are about half an inch long. It is well known that the common form of this plant (var. *macrocarpum*) has fruits about two inches long. This is the variety that has increased so remarkably this century and thereby favoured the increase of the Large Elephant Hawk and Bedstraw Hawk moths. In the Hancock Museum there are thirty-three sheets of *C. angustifolium* of which twenty-eight are var. *brachycarpum*. Most of these were found last century growing wild in craggy localities such as Tweed banks 1831; Swale banks four miles west of Richmond 1831; Invercauld 1832; Hareshaw Linn Bellingham n.d.; Burnmouth ravine n.d.; Craigie Burn n.d.; Derwent banks Blanchland n.d.; above Langleeford n.d. There is one sheet with both varieties from Crag Lough and another with var. *macrocarpum* from near Aldstone, Cumberland collected by John Hancock in 1837. It is clear therefore that both varieties occurred in similar wild localities last century but that var. *brachycarpum* was relatively more common than it is now. This poses the question as to whether the predominant var. *macrocarpum* so well known today is a distinct race with a different origin and ecological preference.

I have described the remarkable increase of the Orange Tip butterfly *Anthocharis cardamines* in *Ent. Record* 91, pp. 16, 42 and 158. Since that article appeared Mr M. E. Braithwaite has reported Orange Tips at Gordon Moss on 10 June 1979, and Mr P. Summers saw them there on 17 June 1979. These are the first records known for Gordon this century. Mr Summers also saw one by the A7 three miles north of Galashiels on 19 June 1979.

Mr W. Davidson of Kelso recorded Orange Tips at Makerston 25 May 1977, Roxburgh Castle 27 May 1977, Lochton 20 May 1978 and Makerston 17 June 1979.

In East Lothian Orange Tips have been seen near Pencaitland 1977–9 by R.W. Barker and P. Summers and at Humbie by Dr G. Waterston 1977–9.

In Northumberland the recovery has similarly been widespread.

With Michael E. Braithwaite (left) at the publication of *The Botanist in Berwickshire* at the Annual General Meeting of the Berwickshire Naturalists' Club, 1990. Photograph by Tweeddale Press.

Males were seen each year 1977–9 at Longframlington by D.G. Burleigh and at Elyhaugh 1977–8 by Mrs P. Henzell.

Another species of butterfly which is increasing its range is the Wall *Lasiommata megera*. According to J.E. Robson it suffered an unaccountable crash last century (in 1861). This was widespread in northern England. This century it has very gradually recovered. In Durham and Northumberland the recovery has been mainly in the coastal area. The records for Northumberland are as follows – 1769 Simonburn (Wallis): Newcastle 1826 (A. and J. Hancock); 1839 Twizel (Selby): 1857 General (Wailes): 1871, Rothbury (Maling); 1870, 1871 Beadnell (Maling); 1880 Spittal (Renton): 1886 Berwick (Bolam); 1891 Norham (Miss Dickinson); 1924 Bamburgh (Harrison); 1929 Corbridge (Cooke); 1976 Wylam and Whitley Bay (Parrack); 1977 Benwell (Eales): 1979 Seaton Sluice and Holywell (Bainbridge and Douglas).

The records for Berwickshire are as follows: 1832 Johnston; 1873 Lauder (Simpson): pre–1902 Billie Mains (Allan): Edrom 1955 (Logan Home). Last year (on 5.8.1978) Mr E. Hamilton recorded in the *Journ. Edin. Nat. Hist. Soc.* p. 31 a party of five or six of these butterflies flying over a grassy cliff near Siccar Point.

A nest of the common wasp *Paravespula vulgaris* was built under the roof of the former police headquarters at the "Kylins" Morpeth. Its dimensions were length 2 ft. (61 cm), depth 1 ft. (30 cm). It was photographed on 14 October when many drones were present on the exterior and again on 21st November when a single dormant young queen was present clinging to the outer shell. This was removed for identification. It had the typical markings of *germanica* on the face and abdomen. This was of interest since the brown light and dark banding as well as the texture of the nest agreed with *P. vulgaris* as shown in the book entitled Wasps by J.P. Spradbury (1973) Plates X and XI. Activity in the nest continued through the mild autumn up to the first frost of 10–11 November. I am indebted to Mr Joseph Gibb of Seaton Sluice for information and photographs of this remarkable nest. Three dead workers brought by Mr Gibb from below the nest had the typical markings of *P. vulgaris*. This suggests that the germanica queen found hibernating on the exterior of the nest was not a daughter of the colony but had found its way into the roof space and came from elsewhere.

The Bearberry *Arctostaphylos uva-ursi* (L) Spreng. Does it still occur in Berwickshire?

In the *Flora of Berwick-upon-Tweed* Vol II published in 1831 Dr George Johnston recorded this plant under the name *Arbutus Uva-ursi* Red Bear-berry "on the west side of Dirrington Law, plentiful. Mr Thomas Brown. June". This station is in the Longformacus 10km. Grid Square NT 65.

The record was repeated in 1845 by Rev. H. Riddell in the New Statistical Account p. 94.

In 1858 Rev. W. Darnell M.A. (President) mentioned in his Anniversary Address that the Club met at Greenlaw on 28th July and on the walk to Black Castle Ring and the Kaimes by way of the "Fawngrass" (Fangrist) burn *Arbutus uva-ursi* was observed along with *Triglochin palustre* and *Pinguicula vulgaris*. Unfortunately the precise location is not stated but it seems unlikely to have been Dirrington Law (*H.B.N.C.* IV, p. 63). In 1886 Dr Charles Stuart recorded the plant as growing plentifully with *Vaccinium vitis-idaea* Cowberry on the west side of Dirrington Law (*Trans. Bot. Edinb.* 6, 80).

In 1932 J.H. Craw wrote of the Bearberry "In former days reported from Dirrington but as it has never been found again this would seem to have been in error." (*H.B.N.C.* XXVIII, 20). However, I now have evidence that the old records were correct since there is a pressed specimen in the R.C. Embleton collection in the Hancock Museum. The label attached states that the collector was Dr G. Johnston, date 1832, site Dirrington Law.

As Bearberry can very easily be confused with Cowberry the following feature should be looked for: In Bearberry the leaves tend to taper towards the stalk i.e. they are more characteristically cuneate than in Cowberry. It would be of interest to see if the Bearberry is still growing there.

REFERENCES

Arnold, C. A. 1939. "Observations on fossil plants from the Devonian of eastern North America. IV. Plant remains from the Catskill delta deposits of Northern Pennsylvania and southern New York". *Univ. Michigan Contrib. Mus. Paleont.*, 5, 271–314.

Barnard, P. D. W., 1960. "*Calathospermum fimbriatum* sp. nov. a Lower Carboniferous Pteridosperm Cupule from Scotland". *Palaeontology* 3, 265–273.

Beck, C. B. 1960. "The identity of *Archaeopteris* and *Callixylon*", *Brittonia* 12, 351–368.

Chaloner, W. G. and Pettitt, J. M. 1964. "A seed megaspore from the Devonian of Canada". *Palaeontology*, 7, 29–36.

Long, A. G. 1960. "On the Structure of *Calymmatotheca kidstonii* Calder (emended) and *Genomosperma latens* gen. et sp. nov. from the Calciferous Sandstone Series of Berwickshire". *Trans. Roy. Soc. Edin.*, 64, 29–44.

Long, A. G. 1968. "Some specimens of *Mazocarpon*, *Achlamydocarpon* and *Cystosporites* from the Lower Carboniferous Rocks of Berwickshire". *Trans. Roy. Soc. Edin.*, 67, 359–372.

Miller, Hugh, 1869. *The Testimony of the Rocks*. Edinburgh, p. 411.

Pettitt, J. M. and Beck, C. B. 1968. "*Archaeosperma arnoldii*, a cupulate seed from the Upper Devonian of North America". *Contrib. Mus. Palaeontology Univ. Mich.* 22, 139–154.

Surange, K. R. 1952. "The morphology of *Stauropteris burntislandica* P. Bertrand and its megasporangium *Bensonites fusiformis* R. Scott". *Phil. Trans. Roy. Soc. London*, 237B, 73–91.

CHAPTER 6

Articles on Fossil Plants

a) Palaeobotanical Reminiscences
b) The Cupule – Carpel Theory
c) The Fossil Plants of Berwickshire

(a) **Palaeobotanical Reminiscences. From the** *History of the Berwickshire Naturalists' Club, Vol. XL, part 3, pp. 179–189. 1976.*

After completing two years of research under Professor W.H. Lang at Manchester University (1937–9) I then decided to enter the teaching profession as I had no means of financial support. I took my teacher's certificate at Hull University College (1939–40) and then offered myself for military service but was turned down on account of my crippled left foot – the result of a shooting accident when I was 14 years of age. My first teaching post at Lewes County School for Boys, East Sussex was only temporary (1940–42) and I moved from there to Leek High School for Boys, North Staffordshire where I remained until the end of 1944. As I was not happy in this school I looked for another post and was successful in being made Assistant Science Master at the Berwickshire High School, Duns, Scotland, where I commenced on 4 January 1945 (the fifteenth anniversary of my shooting accident).

While at Leek I had been able to complete for publication two papers on the coal-ball fossils *Botryopteris hirsuta* Will. and *Lagenostoma ovoides* Will. These appeared in the *Annals of Botany* (Long 1943 and 1944) and I still cherished the desire to describe the leaf-borne buds of *Botryopteris antiqua* discovered in Pettycur material as mentioned in the first of these papers. However, the difficult circumstances I encountered when living in lodgings and frequently changing address made me wellnigh despair of ever doing anything further. I applied for various posts which I felt would have given me the chance to realize my hopes, but to no avail. In my disillusionment I turned to my first love – entomology, and built up gradually an apiary of about thirty stocks of bees, while collecting lepidoptera, trichoptera and other orders of insects.

At first my wife and I lived in lodgings at the foot of Bridgend in Duns but these were temporary. In 1946 we moved to the vacant

Family group (Gladys, David and Jean) at bridge over College Water, near Kilham, Cheviot Hills, Northumberland.

Preston Schoolhouse near the Duns–Grantshouse road. This was our first home and conveniently near the River Whiteadder where I first learned the gentle art of fly fishing. It was while living at Preston that our daughter and son were born in 1947 and 1948. Another noteworthy event was the great flood of 1948 in August. This proved to be of some palaeobotanical significance as it scoured the banks and bed of the Whiteadder and cleaned the shingle beds in readiness for my later discoveries. As if to further the good work the year 1956 brought another flood, this time in September. My fossil collecting, however, did not begin in earnest until 1957. By this time we were living in Gavinton three miles west of Duns near to the Duns to Greenlaw road. We moved there in October 1949 after a record honey year in which my eight stocks of bees averaged 110 pounds of honey per stock. It was my interest in bees which kept me grounded in Berwickshire as once again I was not settled in the school where I was teaching. This state of affairs lasted until 1966 when I reluctantly decided I would have to leave.

My first attempt at collecting fossil plants in Berwickshire was made in 1945. I cycled to the coast near Cockburnspath and dug compressions from shale near Cove Harbour. These I sent to Professor Lang who replied from Withington Manchester, on 20th May 1945 informing me that they were from the Lower Carboniferous. I could not get a geological map of the area at first as stocks had been bombed during

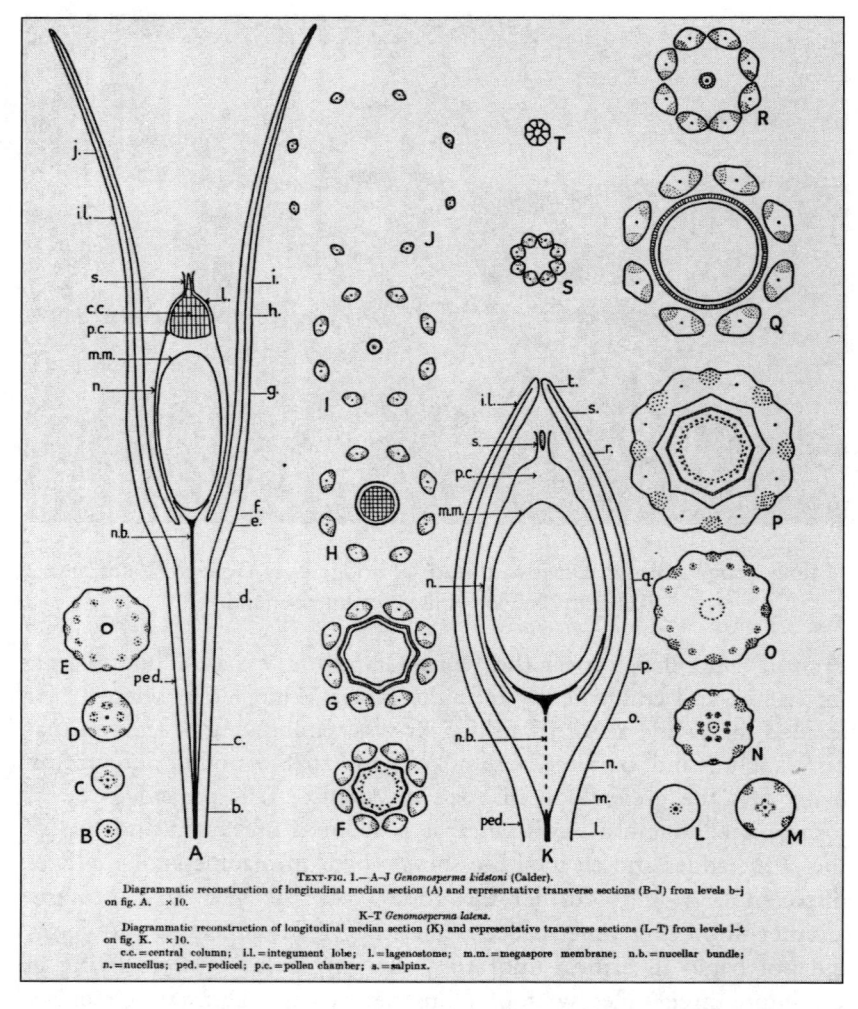

Line diagrams of seeds of the seed-ferns *Genomosperma kidstonii* Calder and *G. latens* Long. Reproduced from *Transactions of the Royal Society of Edinburgh*, volume 70 (1977).
Reproduced by courtesy of the Royal Society of Edinburgh.

the war. Later an old geological map was discovered at school in the geography department and this I copied. About 1949, after moving to Gavinton I resolved to try and continue my work on *Botryopteris antiqua* in some Pettycur material which I had been given at Manchester. I went to Manchester in early 1950 to arrange for the material to be sent up to Duns. I prepared celluloid solutions and made some peels but as I had no cutting facilities the work proved impracticable and I gave up to it. Meanwhile I had tried trout fishing in the Langton Burn

Langton Burn, Berwickshire, from below the ruined Hanna's Bridge, 400 yards NNE of the village of Gavinton. Here the best specimens of fossil seeds of the seed-fern *Genomosperma* were found in loose blocks washed out of the calciferous sandstone above the bridge.

near Gavinton and in 1951 at the unfinished ruined bridge known locally as Hanna's Bridge I got some sandstone blocks containing plant fossils. I casually examined these but dismissed them as useless compressions. Six years later I returned to this spot and made a closer examination with good results as my blocks gave me my best specimens of *Genomosperma (Calymmatotheca) kidstonii* and *G. latens* (Long 1960a). Although I had the two volumes of Scott's *Studies in Fossil Botany* in a book-case in our Gavinton cottage dating back to 1760 I did not realise that the Langton Burn was mentioned on page 141 of volume II where Scott mentioned it as a locality for *Stenomyelon*. My move to Gavinton was thus neither science nor pre-science and could therefore have been only one of two possibilities, either Providence or coincidence.

In fact I was too engrossed in entomological studies at that time to pay much attention to fossil plants. I bought a portable Pioneer generator (ex U.S.A. army equipment) and used it to operate two mercury vapour light traps from an old ex post office Morris van and collected moths and caddis flies in out of the way spots like Kyles Hill, Gordon Moss, Elba, the Retreat near Abbey St. Bathans and the Hungry Snout in the Lammermuirs. Other spots visited were Broomhouse on the Whiteadder, Old Cambus Quarry near Siccar Point, Burnmouth, the Hirsel near Coldstream and Paxton. In this way I scoured the Berwickshire countryside and compiled card indexes of the insects and plants, not to mention ringing and photographing birds.

It was soon after the Suez crisis of 1956 and while preparing to resume my moth collecting that I received a letter dated 17th February 1957 which changed the tenor of my life in a most unexpected way. This letter was from P. D. W. Barnard of Birkbeck College, London, and at that time a complete stranger to me. His letter acted like a catalyst and spurred me to make a search of the Langton Burn near Gavinton. In this letter he wrote,

> I have been reading through literature describing the Calciferous Sandstone plants. Two localities from which material is recorded are near to your address which I have just come across in the Palaeobotanical Report, 1954–6. These localities from which petrified remains have been described are LANGTON BURN, 400 yards north of Gavinton, and Edrom. If you have collected from any of these localities recently or know of any other localities where material has been collected recently I should be pleased to receive any extra information concerning this area you can offer.

As a result of this letter I first searched the Langton Burn upstream to Langton Glen but without success. Then I decided to search

Dr Peter D. W. Barnard with his father and mother at The Green, Gavinton, Berwickshire, 1958.

downstream from the road bridge near the Red Brae below Langton Church. When I reached Hanna's Bridge I remembered and re-discovered the blocks first seen in 1951 when fishing. As soon as I saw the first specimens of *Calymmatotheca kidstonii* Calder "the penny dropped" and immediately I realized that here was something of great interest and significance. *Lagenostoma ovoides* from the Upper Carboniferous coal-balls has an integument with eight chambers at its apex obviously suggesting that the integument had evolved by union of eight free lobes. *G. kidstonii* and *G. latens* were making no secret of how this had happened. The sudden revelation of these fossils now gave the moths and bees some respite. I continued to use a garden moth trap but suspended other night collecting and completed my lists of the Berwickshire Macro-Lepidoptera and Trichoptera as best I could. They are published in the *History of the Berwickshire Naturalists' Club*, volumes 34–38. My bees were sold in 1958 and this cleared my feet to proceed with the fossil work. At first I tried making peels by the old celluloid solution method but my attention was drawn to the new technique using cellulose acetate film (Joy, Willis and Lacey 1956). This technique came just in time for my needs. At first I ground blocks down from the surface and used a bench in the garden. Later when I had sold my honey extractor and other gear I used a washhouse which I had rebuilt as a honey-house; this became my laboratory. Besides the Langton Burn material which gave me *Genomosperma* and *Lyrasperma* I soon discovered a remarkable block (in September 1957) on a shingle bed by the Whiteadder near Hutton Mill. This gave me *Stamnostoma* (Long 1960). It was a great thrill to find this exquisite seed and to discover a pair of cupules of which one conveniently contained a single seed. These specimens of *Stamnostoma* cupules have never been bettered though during the recent fine summer (1976) further specimens containing seeds have been obtained from the Wooler Water. These are the first yet discovered in Northumberland. I made over 900 peels from the Hutton Mill block and in all the time taken there was never a dull moment. It produced in addition the new seeds *Hydrasperma* and *Deltasperma*. As *Deltasperma* like *Lyrasperma* possesses only two integumental lobes it became clear that the ovule integument most probably evolved from sterile telomes rather than from a cluster or synangium of fertile telomes as postulated in Benson's theory.

At the time of writing (1977) I have described fourteen species of petrified fossil seeds from the Lower Carboniferous Cementstone Group of S.E. Scotland. These are listed below.

Order LAGENOSTOMALES
Family GENOMOSPERMACEAE
Genomosperma kidstonii (Calder) 1960
Genomosperma latens 1960
Family EOSPERMACEAE
Eosperma edromense 1966
Deltasperma fouldenense 1961
Eccroustosperma langtonense 1961
Camptosperma berniciense 1961
Family EURYSTOMACEAE
Lyrasperma scotica (Calder) 1960
Eurystoma angulare 1960
Eurystoma burnense, 1966, 1969, 1975
Tantallosperma setigera Barnard and Long 1973
Dolichosperma sexangulatum 1961, 1975
Dolichosperma pentagonum 1975
Family LAGENOSTOMACEAE
Stamnostoma huttonense 1960 and *S. bifrons*, 1961 probably one species
Order CARDIOCARPALES
Mitrospermum berwickense (in press) now *M. bulbosum*

It is my opinion (not fully demonstrable) that *Genomosperma* seeds are assignable to *Rhetinangium*.

Eosperma seeds may be assignable to *Aneimites*;
Eurystoma and *Lyrasperma* seeds may be assignable to *Calamopitys*, *Stenomyelon* and *Alcicornopteris*:
Salpingostoma may be assignable to *Calathopteris*.
Stamnostoma is assignable to *Pitus*. I stress that these opinions are suppositions based partly on incomplete evidence such as association and only new specimens showing proof of connection can demonstrate what is the truth.

Among these fourteen species of fossil seeds is one which I consider of unusual significance though my deductions must not be taken as any more than theoretical. I refer to *Camptosperma berniciense*. This seed I have placed in the Eospermaceae along with *Eosperma*, *Eccroustosperma* and *Deltasperma*. They are all obviously related and show a transition from bilateral symmetry to dorsiventral campylotropy. In its campylotropous condition *Camptosperma* shows a striking resemblance to that in some Angiosperm seeds. In 1966 I pointed out that this could shed light on the origin of the second integument in many Angiosperm seeds. In *Camptosperma* the curvature of the seed has led to the development of a pronounced dorsal lip or hood hanging over the mouth of the seed as if to protect the pollination droplet. At the same time the two short free lobes of the integument (possibly secretory) have sunk downwards and inwards alongside the lagenostome somewhat like the two paws of a squirrel when holding a nut to eat. Here we have possibly an inception of a division of function and structure within the single integument of this primitive seed. It would only need the two free distal lobes of the integument to unite around the lagenostome below the hood-like dorsal lip to produce a condition like that seen in those Angiosperm ovules possessing outer and inner integuments. The fact that in these the outer integument is absent ventrally (between seed-body and funicle) supports this view that the outer integument in Angiosperm ovules could really represent a lateral and forward overgrowth of the first integument correlated with curvature and serving to protect the pollination droplet from rain. (N.B. this theory was later abandoned in favour of a cupular origin of the second or outer integument.)

Besides contributing to a theory of the evolution of ovules the Berwickshire fossils have pointed the way to what I call the cupule-carpel theory which, if true, sheds light on the origin of Angiosperms. It was in the Hutton Mill block found in Autumn 1957 that I first discovered attached seeds of *Eurystoma angulare* (Long 1960c and 1965). These were borne in a primitive reflexed cupulate organ obviously composed of cylindrical non-laminar branchlets produced by cruciate dichotomies. It suggested to me that an Angiosperm carpel probably represents – not a conduplicate laminar leaf – but a bivalved cupule originally constructed of cylindrical branchlets of an appendicular organ comparable to a leaf. With the discovery of cupules of *Calathospermum fimbriatum* Barnard at Oxroad Bay and in the bed of the Whiteadder it has become evident that the typical Angiosperm carpel may have evolved from a pair of cupules derived from an entire dichotomous frond originally composed of cylindrical non-planated branchlets. These two cupules must have later coalesced to form the two halves or valves of a single cupule or carpel.

The small seeds named *Hydrasperma* were also found in the aforementioned Hutton Mill block and described in 1961. No more came to light until 1971 when sixteen tiny attached ovules were discovered inside a pair of cupules from Oxroad Bay on the East Lothian coast. These paired cupules are borne on a very short common stalk (originally on a cylindrical axis) and each is itself bivalved and has a proximal umbrella-like region from which tapering cylindrical lobes project like fingers from the palm of one's hand. Ovules are sessile and pendent

The late Professor John Walton of Glasgow University.
From a portrait by Alberto Morrocco, 1962.

below the umbrella region of the cupule so that placentation is laminar. This supports the view that in carpels laminar placentation is primitive. A further interesting feature of this pair of cupules is that one has eight microsporangia. This suggests that Pteridosperm sporophylls were originally bisporangiate i.e. capable of bearing both micro-and megasporangia. One question raised but not answered by the *Hydrasperma* cupules is whether or not Angiosperms have two kinds of carpels evolved slightly differently. Type 1. the *Calathospermum* type having two halves each of which represents one half of a complete megasporophyll (cf. *Magnolia*). Type 2. the *Hydrasperma* type, in which one megasporophyll produced a pair of carpels united basally and each possessing two valves representing two quarters of the complete megasporophyll. Such a type would have its carpels in pairs probably forming bicarpellary gynoecia as in *Salix*. The arrangement would then be a spiral of pairs, rather than of unit carpels.

During the early part of my work on the fossil flora of the Cementstone Group, I contacted the late Professor John Walton of Glasgow University and visited him during December 1957. He kindly communicated my earlier papers to the Royal Society of Edinburgh and gave me much support in many other ways and this continued right up to the time of his death in early 1971. It was through Professor Walton that I learned of D. H. Scott's correspondence in connection with his search for a portrait of Henry Witham. Later I was able to publish this correspondence in the *History of the Berwickshire Naturalists' Club*, Vol. 34, pp. 267–273 (1959). Since coming to the Hancock Museum in 1966 I have discovered in the Museum a fine petrological section which I recognized immediately as one figured by Henry Witham in his book *Observations on Fossil Vegetables* 1831, Plate III, fig. 1. It is a transverse section of the Lennel Braes tree *Pitus antiqua* and shows two areas of preserved tracheids surrounded by the characteristic honeycomb appearance of the secondary wood where it has been disrupted by crystallisation of the mineral matrix.

Later (in 1968) this Museum was given a collection of 111 old petrological slides by Ushaw College. Many of these slides had been inscribed by means of a diamond with the names of places of origin. The principal localities are Tweed Mill; Allanbank, Berwickshire; Craigleith (1830 and 1831); Scarbro; Ushaw; Gateshead; Rothbury and near Glasgow. It is clear that these slides must have been Henry Witham's. Probably they would be given to Ushaw College by Henry Witham's son who was a Roman Catholic priest and lived to the age of 100. One of the slides from Craigleith is the section of the branch of *Pitus withami*, which fell off the tree trunk when it was blasted from the rock in Craigleith Quarry, Edinburgh. This Quarry is at present (1977) being filled in. A section of this branch was figured by Witham in the *Transactions of the Natural History Society of Northumberland, Durham and Newcastle upon Tyne* (1831), Vol. 1, Plate XXV, fig 5 and in Witham's second book *The Internal Structure of Fossil Vegetables* 1833, Plate VI, fig. 5. It was first named *Pinites medullaris* by Lindley and Hutton in their *Fossil Flora* Vol. I., pp. 13–14. The generic name *Pitus* was first applied to *Pitus antiqua* from Lennel Braes near Coldstream on Tweed. In the same work Witham also described *Pitus primaeva* from Tweed Mill about a mile below Lennel Braes. Although *Pitus antiqua* is thus the type species of the genus, the Craigleith species (*Pitus withami*) was described first but without a scientific name. The first specimen was obtained in 1826 and described in the *Philosophical*

Magazine for Jan. 1830, Series 2, Vol. VII, pp. 23–31 'On the vegetation of the First Period of an Ancient World'. This stem was 36 feet long and 3 feet in diameter at its base. A second specimen was discovered at Craigleith in 1830 and described by Witham in the *Transactions of the Natural History Society of Northumberland, Durham and Newcastle upon Tyne*, Vol. I, pp. 294–301 in 1831. This description was also without a scientific name. In this paper Witham also mentioned a third specimen from Craigleith on p. 297 and said that it was blasted, during which a branch broke off. This was the branch sectioned and figured by Witham in 1833. The specimen, however, was earlier figured by Lindley and Hutton in 1831 in their *Fossil Flora* and given the name *Pinites medullaris* whereas the tree from which the branch fell was named *Pinites withami*. This identity of *P. withami* and *P. medullaris* was first pointed out by Witham in 1833 and later confirmed by D.H. Scott in 1902. The generic name *Pinites* was first used by Lindley and Hutton in 1831 for the type species *Pinites brandlingi* (*Fossil Flora*, Vol. 1, p. 1 and Plate 1). This specimen also referred to as the Wideopen Tree, was Upper Carboniferous in age and came from Wideopen, 5 miles north of Newcastle upon Tyne. It was described and figured (but not named) also by Witham in 1833, p. 31, and Plate, 4 figs. 1–5.

It was later (1917) placed in the Genus *Dadoxylon* and is now generally recognized as identical

Trunk of *Pitus withami* from Craigleith Quarry, erected at the Royal Botanic Garden, Edinburgh. This is Specimen 2 of Witham found at Craigleith Quarry, Edinburgh, in 1830. Photograph by Royal Botanic Garden, Edinburgh, and published in *Transactions of the Royal Society of Edinburgh* (1979). Reproduced by courtesy of the Royal Society of Edinburgh.

with *Cordaites* Unger 1850. Hence Witham's name *Pitus* has superseded the name *Pinites* as far as the Berwickshire and Craigleith fossils are concerned and the specific epithet *medullaris* has been dropped as a synonym for *withami*.

It was the late Professor W.T. Gordon who first showed in 1935 that *Pitus* sometimes has attached leaves borne fairly close together near branch apices. The leaves probably became separated by stem growth and possibly reflexed. The stem cortex seems to have been relatively soft so that the leaves readily became detached without leaving attached petiole bases. Each petiole is swollen at the base. The leaf trace is at first surrounded with secondary xylem. On passing through the stem cortex it divides into three and then six vascular bundles which join in the petiole base to form a large V or U- shaped bundle in cross section. Such a large bundle is not unlike that of *Lyginopteris* and suggests that the frond had a large transpiring surface. Detached petioles of this type occur mixed with stems of *Pitus primaeva* in Berwickshire and are know as *Lyginorachis papilio* Kidston. Gordon's attached petioles were probably too young to have a fully expanded frond so he interpreted the leaves as phyllodes somewhat like the leaves of certain *Araucarias*. *The size, form and angle of bifurcation shown by Lyginorachis papilio (detached) suggest that it was a typical Pteridosperm frond up to one or two feet in length and thus comparable in size to the compression known as Sphenopteris affinis,* a very common 'fern' frond in the Lower Carboniferous oil shales (see frontispiece to Hugh Miller's *Testimony of the Rocks*).

Some petioles of *Lyginorachis papilio* were trifurcate i.e. they had a median stem-like third rachis (*Tristichia ovensi* Long), between the two outer secondary rachises of the Y-shaped frond. In the compression *Diplopteridium teilianum* Walton the median rachis bore branches which in turn branched by repeated wide dichotomies ultimately bearing clusters of terminal micro-sporangia (*Telangium*). Evidence from the Crooked Burn compressions suggests that other trifurcate fronds probably of *L. papilio* bore seed cupules of the species *Stamnostoma huttonense* Long. If the evidence is interpreted correctly it suggests that *Pitus* was an arborescent Pteridosperm bearing three kinds of frond comparable in size to *Sphenopteris affinis*. No doubt other such arborescent Pteridosperms existed in Lower and Upper Carboniferous times though the majority seem to have been of shrub habit.

Henry Witham was a founder member of the Natural History Society which built and for long maintained the Hancock Museum. I feel sure he would have rejoiced to see the branch of science he so

Excavating for compressed fossil plants in the shales at the Crooked Burn near Foulden Newton, Berwickshire. A. G. Long (left) with I. McWhan, about 1963/4. Photograph by T. Huxley, Nature Conservancy, Edinburgh.

famously pioneered being continued and furthered in the institution he helped to launch 148 years ago.

Strange and wonderful are the ways of Providence. "O the depth of the riches both of the wisdom and the knowledge of God! How unsearchable are his judgements, and his ways past tracing out" (Romans 11:33).

(b) **The Cupule-Carpel Theory. A Defence. From** *Trans. Bot. Soc. Edinb. 44, 281–285*

In his recent book *Palaeobotany and the Evolution of Plants* (1983) Wilson N. Stewart has dealt with the theory of Angiosperm evolution from Pteridosperm ancestors and has discussed the possible evolution of Angiosperm carpels from Pteridosperm cupules and of bitegmic anatropous ovules from unitegmic orthotropous ovules. His account of my version of this theory does less than justice to what I have written (Long, 1966, 1975, 1977a).

On p. 386 Stewart has written,

> The origin of the carpel from those Paleozoic pteridosperms forming cupules is supported by Delevoryas (1962), Andrews (1963) and Long (1966). Delevoryas favours lyginopterid Pteridosperms of the *Calathospermum* and *Gnetopsis* type where the cupule bears several orthotropous ovules in the base . . . If as suspected, a cupule of *Calathospermum* is a modified frond, then the model suggested by Delevoryas provides us with a carpel that is foliar, and this is in keeping with the presumed foliar nature of the Angiosperm carpel. Long (1966, 1967), on the other hand favours those lyginopterid Pteridosperms where the cupule is clearly only a segment of a frond. This offers difficulties because the Angiosperm carpel is usually homologized with an entire leaf. Other problems are the orthotropous unitegmic ovules of lyginopterid Pteridosperms that are unlike the anatropous, bitegmic ovules of Angiosperms. In homologizing the lyginopterid cupule with the ovary wall, Long is required to explain the origin of the second integument as an outgrowth of the first integument or the chalazal stalk of the ovule. Except for the model presented by Stebbins (1986) no others attempt to explain the origin of the anatropous bitegmic ovule.

This statement leaves much to be desired as an interpretation of the cupule-carpel theory I published. For example in 1977 I completely revised my theory of the origin of the second ovular integument pointing out three objections to my 1966 theory (see Long 1977a, p. 24). This revision is not mentioned by Stewart although the paper is quoted. My published work also included references to the *Calathospermum* cupule (on which I have worked) and describes new facts from a new locality (the River Whiteadder in Berwickshire). From these facts described and illustrated in my 1975 paper (e.g. Text-figs. 4–6) I reinterpreted the *Calathospermum* cupule as derived from twin cupules so that the innermost "central lobes are in reality lateral adjacent lobes of two cupules, and ovules are laterally placed and not basal originally". Even in my 1966 paper I had shown that I was aware that a cupule was similar to a foliage leaf but my interest was to show that primitive cupules were not necessarily laminate and the appearance of being conduplicate could, in its origin, have been derived from dichotomy of terete mesomes and telomes in successive planes at right angles. The words I used (Long 1966, p. 370) were,

> If the cupule theory of the carpel is accepted this does not overthrow the view that the carpel is appendicular and therefore comparable to a foliage leaf but rather it overthrows the view that the carpel was originally laminar and secondarily folded along its midrib. It therefore invalidates the idea that a carpel was a conduplicate leaf. In its place it puts the idea that the carpel was a system of terete dichotomous telomes and mesomes having dorsiventral vascular bundles but no lamina. The primitive structure was therefore one of a simple branch system in which the branches occurred in successive planes at right angles. Adnation is conceived of as having occurred basally as a result of hormone suppression of branching. On this view the carpel never was a laminar leaf.

In the discussion of my 1975 paper I wrote (p. 287)

> If a *Calathospermum* cupule is the product of fusion of two cupules this modifies the above theory (that of Long 1966) so that it has to be re-written as follows; The Angiosperm carpel could have evolved from a single dorsiventral bilateral cupule which in turn had evolved from a pair of cupules radially constructed and borne on the two divisions of the bifurcated rachis. Each valve of the carpel on this theory is the representative of a single cupule originally of radial construction.

It is obvious therefore that this in part corroborates Thomas's theory (Thomas 1931).

These quotations are witness to my awareness and agreement with Delevoryas that the cupule of *Calathospermum* could be regarded as a precursor of an Angiosperm carpel and was an appendicular organ like a leaf. In 1977, however, I carried the cupule-carpel theory further after consideration of the cupules of *Hydrasperma* which I identified with the compression *Sphenopteris bifida* L. & H. described originally from Burdiehouse (Long 1977b). The cupules were found to be of two kinds (a) some with two valves and (b) some paired on one stalk, each with two valves. Those of group (a) formed by single dichotomy

I called mega-cupules. Those of group (b) formed by double dichotomy I called hemi-cupules. *Calathospermum* on this interpretation is a mega-cupule formed by union of twin cupules. From this interpretation I inferred that carpels likewise may be mega-carpels, occurring singly, and derived by single dichotomy of a megasporophyll. On this interpretation *Magnolia* is cited as a possible example of a mega-carpel. In contrast, if double dichotomy of a megasporophyll occurred union of the quarters in two pairs would form two hemi-carpels e.g. *Salix*.

Fossil evidence that carpels are sometimes borne in pairs was cited from Krassilov (1973) who described the Mesozoic *Trochodendrocarpus arcticum* (Heer) Kryst. In this fossil spirally arranged seed-pods are borne either in pairs or singly and each pod can be interpreted as paired hemi-carpels formed by double dichotomy of a single megasporophyll while the single pods can be interpreted as megacarpels derived by single dichotomy of a megasporophyll.

Concerning anatropous ovules and the second integument of Angiosperm ovules I wrote (Long 1977a, p. 24)

> In *Hydrasperma* the cupule lobes show a definite tendency to curve inwards. In some positions ovules were close to sterile cupule lobes and it would not be surprising if such a curved lobe sometimes became fused to an ovule. Such a theoretical derivation of the outer integument from one or more cupule lobes helps to explain the histological difference between outer and inner integuments of angiosperm ovules.

Thus I concluded (p. 26) "that the outer integument has most probably evolved from one or more sterile cupule lobes sometimes branched and closely associated with the ovule". The rest of the cupule, proximal to the lobes, I consider has become the carpel proper by basal webbing so that the ovules are positioned at the new margin but facing outwards by inversion.

Further to this extension of the cupule-carpel theory in 1977 I have also pointed out that the fertile median 'rachis' borne in the fork of a fertile Pteridosperm frond may be re-interpreted as a stem borne on a leaf, (Long 1976a, and cf. *Botryopteris*, Long 1943). This does not imply adnation of stem and leaf as in Melville's gonophyll theory but something much more fundamental viz. that stem and leaf were originally one. Therefore a stem borne on a leaf, or a flower subtended by a bract are vestiges of unity. Such primitive unity thus underlines present diversity, stem, leaf, and cupule or carpel having originated in a common telomic truss. Each organ appears to have evolved functionally a leaf for photosynthesis, a receptacle for support and cupule or carpel for protection and nourishment of ovules and seeds. Hence the axillary branch bearing seed cupules and/or microsporangia in the angle of a Pteridosperm frond could represent a median member of a trifurcate appendicular branch such as occurs in *Trimerophytopsida* or ferns and their allies e.g. *Psalixochlaena berwickense* (Long 1976b). If this view is valid it would mean that the Pteridosperms and Angiosperms go back probably through *progymnosperms* such as Aneurophytales and Archaeopteridales to Trimerophytopsida and through them to the Rhyniopsida.

Palaeobotanical evidence supports the view that ferns and Pteridosperms show a similar parallel evolution but that Angiosperms most probably came from Pteridosperms. The cupule-carpel theory supports this but the difference in axillary branching of *Calamopitys* Galtier and Holmes, (1982) reminds us that as yet the fossil record has not given Angiosperm ancestors themselves but it has yielded clues as to where their ancient relatives are to be found.

REFERENCES

Andrews, H. N. (1963). "Early Seed Plants". *Science*, 142, 925–931.

Delevoryas, T. (1962). *Morphology and Evolution of Fossil plants*. Holt, Rineholt and Wilson, New York.

Galtier, J. and Holmes, J. C. (1982). "New Observations on the Branching of Carboniferous Ferns and Pteridosperms". *Annals of Botany*, 49, 737–746.

Krassilov, V. (1973). "Mesozoic plants and the problem of angiosperm ancestry" *Lethaia* 6, 163–178.

Long, A. G. (1943). "On the occurrence of buds on the leaves of *Botryoperis hirsuta* Will". *Annals of Botany*, N.S.7, 133–146.

Long, A. G. (1966). "Some Lower Carboniferous fructifications from Berwickshire, together with a theoretical account of the evolution of ovules, cupules and carpels". *Trans. Roy. Soc. Edinb.* 66, 345–375.

Long, A. G. (1975). "Further Observations on some Lower Carboniferous Seeds and cupules". *Trans. Roy. Soc. Edinb.* 69, 267–293.

Long, A. G. (1976a). "*Calathopteris heterophylla* gen. et sp. nov., a Lower Carboniferous Pteridosperm bearing two kinds of petioles". *Trans. Roy Soc. Edinb.* 69, 327–336.

Long, A. G. (1976b). "*Rowleya trifurcata* gen. et sp. nov. a Simple Petrified Vascular Plant from the Lower Coal Measures (Westphalian A) of Lancashire". *Trans. Roy. Soc. Edinb.* 69, 467–481.

Long, A. G. (1976c). "*Psalixochlaena berwickense* sp. nov., a Lower Carboniferous Fern from Berwickshire". *Trans. Roy. Soc. Edinb.*, 69, 513–521.

Long, A. G. (1977a). "Lower Carboniferous pteridosperms and the origin of angiosperms", *Trans. Roy. Soc. Edinb.*, 79, 13–35.

Long, A. G. (1977b). "The resemblance between the Lower Carboniferous

Sphenopteris bifida Lindley and Hutton and *Hydrasperma*". *Trans. Roy. Soc. Edin.*, 70, 129–137.

Stebbins, G. L. (1976). "Seeds, seedlings and the origin of angiosperms". In *Origin and Early Evolution of Angiosperms*. (Ed. Beck, C.B.) Columbia University Press. New York pp. 300–311.

Stewart, W. M. 1983. *Palaeobotany and the Evolution of Plants*. Cambridge University Press.

Thomas, H. H. (1931). "The Early Evolution of the Angiosperms". *Annals of Botany*, 45, 646.

(c) **The Fossil Plants of Berwickshire – a review of past work.**

Part 1. Work done during the last century.

Introduction. From *History of the Berwickshire Naturalists' Club*, Vol. XXXIV, pp. 248–273.

The purpose of this paper is to bring together known records of fossil plants hitherto found in Berwickshire with the object of reviewing the discoveries already made, so as to assist any future worker to obtain such information without the labour of searching out old records. It should also serve as a guide to the most likely localities from which new specimens of fossil plants may yet be obtainable.

In Berwickshire three major geological formations are represented viz. the Silurian, Devonian and Carboniferous though much of the area is covered with more recent (Quaternary) glacial drift deposits. These three formations are all grouped in the Palaeozoic division of the geological system of classification, the Silurian being the oldest and lying for the most part to the north of the county, where it has been thrust up and laid bare by denudation of the other two formations.

The Devonian, or Old Red Sandstone, lies chiefly on the flanks of the Lammermuirs, though a tongue strikes right through by way of the Upper Monynut to Dunbar (cf. Stevenson, 1850).

The Carboniferous rocks occupy the lower ground of the Merse, roughly forming a triangle between Kelso, Duns and Berwick-upon-Tweed. There are no plant fossils known to the writer which have been found in the Silurian formation of Berwickshire. This is not surprising as the rocks are of marine origin and very ancient (estimated at 440 million years).

The Old Red Sandstone, though of fresh-water origin, is also relatively barren of plant remains in this area. In his book, *The Geology of Eastern Berwickshire*, Sir Archibald Geikie mentioned the occurrence of poor plant impressions in the Reston area. Thus he wrote: "A little above East Reston Mill, some green and pinkish-coloured grits occur on the side of the stream (R. Eye). They are, in some of their bands, very ashy in composition, and contain occasional green shaly beds, in which I detected a few imperfect plants" (p. 26). Geikie again mentioned these fragmentary plants in the concluding paragraph of Chapter III, (Geikie 1863 (a) p. 28).

A more interesting reference to a Devonian fossil plant occurring in Berwickshire is made by Hugh Miller in "The Testimony of the Rocks". This concerns *Archaeopteris (Cyclopteris) hibernica* and is found in his eleventh lecture which was read at The British Association Meeting, held at Glasgow in 1855, under the title "On the Less-known Fossil Floras of Scotland". Concerning this fossil-plant Miller wrote,

> I owe my specimen to Mr John Stewart of Edinburgh who laid it open in a micaceous red Sandstone in the quarry of Prestonhaugh near Dunse, where it is associated with some of the better known icthyic remains of the Upper Old Red Sandstone, such as *Pterichthys major* and *Holoptychius nobilissimus*. Existing as a deep red film in the rock with a tolerably well defined outline, but without trace of the characteristic venation on which the fossil botanist in dealing with the ferns, founds his generic distinction. I could only determine that it was either a *Cyclopteris* or a *Neuropteris*. My collection was visited, however, by the late lamented Edward Forbes only a few weeks before his death and he at once recognized in my Berwickshire fern, so unequivocally an organism of the Upper Old Red Sandstone, the *Cyclopteris hibernicus* of those largely developed beds of yellow sandstone which form so marked a feature in the geology of the south of Ireland, and whose true place whether as Upper Old Red or Lower Carboniferous, has been the subject of so much controversy. (Miller 1869, p. 411.)

It would be of great interest if the presence of *Archaeopteris* at Prestonhaugh could be confirmed by further specimens, as the plant seems to be extremely scarce in the Scottish Old Red Sandstone.

It is in the Carboniferous rocks, however, that the most interesting fossil plants have been discovered in Berwickshire. Some of these come from the lowest beds of the formation, viz. the Cementstone Group of the Calciferous Sandstone Series. These rocks were formed under conditions which occasionally favoured plant petrifaction, so preserving at least part of their internal structure.

These conditions may have been connected with the great amount of volcanic activity occurring at the close of the Devonian period. Thus the volcanic lavas known as the Kelso traps, which are well developed between Kelso and Greenlaw are usually considered to mark the base of the Carboniferous system in that area (cf. MacGregor and

Eckford, 1948 (b), p. 239). In his account of the Lower Carboniferous rocks of Berwickshire Pringle stated that "In this area flows of olivine basalt occur at the base of the Cementstone Group and can be traced from Duns southwards to Greenlaw, and thence across the Tweed west of Kelso to the North flank of the Cheviots. Close above the lavas lies the well known Carham Stone, a cherty magnesian limestone of chemical origin, which was deposited in thin layers in pools subject to desiccation. It is thought that the lime content of these waters has been increased by showers of volcanic dust during the final stages of the Kelso eruptions". (Pringle, 1948 (a) p. 60.)

It is probable that the plants preserved in these lowest strata of the Carboniferous rocks were survivors from the Devonian period, since there was no time-break between the two formations. This is well shown at the coast NW of Pease Bay, where there is an unbroken succession of rocks exposed in the cliffs from the Upper Old Red Sandstone to the lowest beds of the Cementstone Group. Full details of the succession are in the *Geology of East Lothian* by C.T. Clough and others (Clough, 1910, p. 44). Here Clough stated, "The conglomerate with fish and plant remains occurs more than fifty feet above the base of the group." Pringle stated, "The base is taken below a bed of fragments of cementstone and is overlain by a calcareous rock crowded with plant remains, ostracods and Lamellibranchs." (Pringle 1948 (a) p. 61). Similar beds representing the basal portion of the Carboniferous system are also cut by the Whiteadder between Preston Bridge and the Tweed, thus opportunity is afforded in Berwickshire of obtaining some of the most primitive types of Carboniferous plants. It is chiefly this fact which gives so much interest to the pursuit of palaeobotany in the County. In consequence Berwickshire can claim a long connection with the work of well-known palaeobotanists of whom the first was Henry Witham of Lartington Hall in North Yorkshire, - the pioneer investigator of the internal structure of fossil plants.

Henry Witham (1779–1844) and *The Internal Structure of Fossil Vegetables*.

In 1832, the year after the Berwickshire Naturalists' Club was formed Henry T.M. Witham F.G.S., F.R.S.E. published his book *Observations on Fossil Vegetables accompanied by Representations of their Internal Structure as seen through the Microscope.*

This was republished in 1833 under the title *The Internal Structure of Fossil Vegetables found in the Carboniferous and Oolitic Deposits of Great Britain, described and illustrated.* Witham is rightly regarded as the pioneer investigator of the internal structure of fossil plants, though he acknowledged his indebtedness to William Nicol of Edinburgh for the method of preparing thin rock sections. Nicol in turn acknowledged that his method of making sections was a development of that used by Mr Sanderson, lapidary, (Nicol, 1834). In a footnote to this paper the Editor of the Journal, Professor Robert Jameson stated: "This mode of showing the structure of fossil woods has been long known and for years I have been in the practice of recommending it to the attention of geologists". Professor J. W. Judd in a letter to D.H. Scott (see appendix) claimed that Sorby also used the method, and learned it from W. C. Williamson. MacGregor and Eckford stated that Nicol "had merely improved on the methods of preparing transparent sections (of fossil wood and coal) which were already in use by George Sanderson, an Edinburgh optician and lapidary and an early fellow of the Edinburgh Geological Society. Later in the decade, 1850–60, Nicol's prism and his thin section technique were applied by Sorby to the microscopic study of rocks. Thus Witham's discovery of plant petrifactions in Berwickshire led up to a revolution in petrology as well as in fossil botany." (MacGregor and Eckford, 1948 (b), pp. 237–8).

Details concerning Witham's life were collected by Dr D.H. Scott F.R.S., for his Presidential Address to the Linnaean Society in 1911 (Scott, 1911 pp. 17–29). These were later incorporated into a lecture on William Crawford Williamson (1816–95) which was published in *Makers of British Botany* edited by F. W. Oliver, in 1913 (Scott, 1913, pp. 243–60). Dr Scott wrote to several people early in 1911 enquiring for particulars about Witham's life and to find out if any portrait of him was still in existence. Among those to whom he wrote, replies are preserved from Professor J. W. Judd, Sir A. C. Seward, Mr Philip Witham, Professor Isaac Bayley Balfour and Mr G. S. Boulger, together with a letter to Professor Balfour from Captain Norman of Berwick-upon-Tweed. Captain Norman in turn wrote to Rev. I Stark (Former Roman Catholic priest in Berwick) and also to Mr Philip Witham.

All these letters were kept by Dr Scott, along with a copy of Henry Witham's portrait which is published in *Makers of British Botany*, and which was obtained through Mr Philip Witham as a result of these enquiries. After the death of Dr Scott, these documents were given to Professor John Walton of Glasgow University by Dr Scott's daughter, Mrs V. G. Wiltshire. I am indebted to Professor Walton for the loan of these letters and to Mrs Wiltshire for permission to quote them, in order to fill a gap in our knowledge of Henry Witham, whose connection with Berwickshire makes it very desirable that something should be placed on record in the Club's *History* regarding his work on the famous fossil trees which he obtained within the County. In addition I am indebted to the Syndics of the Cambridge University Press for permission to produce the portrait of Henry Witham from *Makers of British Botany*.

As will be seen from Dr Scott's correspondence appended to this paper Henry Witham's original name was Silvertop. He was born in 1779 and was the second son of John Silvertop of Minster Acres, Northumberland by Catherine daughter of Sir Henry Lawson of Brough, Yorkshire. As Henry Silvertop he inherited the Lartington property. He married Miss Eliza Witham, niece and co-heiress of William Witham of Cliffe, Yorkshire, and by Sign Manual took the name and Arms of Witham. Later he became the first Roman Catholic High Sheriff of the County of Durham. He died on 28th November 1844 at the age of 65.

Dr Scott stated that Witham's work on fossil plants belonged to a short period of his life when he was about fifty. What brought Witham to Berwickshire I have never discovered, but as a gentleman of means he would have often traversed the County on his way to and from Edinburgh, where he was a member of the Wernerian Society and of the Royal Society. Thus he read a paper to the Wernerian Society on 5th December 1829, "On the Vegetation of the First Period of an Ancient World" (Witham 1830 (a) pp. 28–29).

Witham was also a member of the Natural History Society of Northumberland, Durham and Newcastle upon Tyne, and was acquainted with the geologist N. J. Winch, who published papers on the geology of Northumberland and the valley of the Tweed (Winch, 1816 and 1831 (e)). Dr Scott mentioned that a letter from Witham to Winch, dated 23rd December 1829, is preserved among Winch's correspondence in the Library of the Linnaean Society.

In Vol. I of Lindley and Hutton's *Fossil Flora of Great Britain* (1831–33) Witham's name appears as a subscriber, and his own work *The Internal Structure of Fossil Vegetables* was dedicated to William Hutton. Dr Scott has quoted a passage from this dedication to illustrate the spirit in which Witham undertook his work: "To lend my aid in bringing from their obscure repositories the ancient records of a former state of things with the view of disclosing the early and mysterious operations of the Great Author of all created things, will ever be to me a source of unalloyed pleasure." Only one Berwickshire fossil plant is mentioned by Lindley and Hutton in their Fossil Flora of Great Britain. This was *Anabathra pulcherrima* (Lindley and Hutton Vol. III, p. 48) which the authors compared to a specimen of *Stigmaria ficoides* with structure preserved. Witham stated that the fossil came from Allanbank, and his account of the plant-remains at this locality is worth quoting: "At Allanbank in

Berwickshire we find shale exposed, containing large quantities of stems of fossil trees, many of which seem to have decayed and to have subsequently been filled with fragments of various vegetables . . . This locality is about 7 miles from Lennel Braes and is near the junction of the Whiteadder and Blackadder. It affords a considerable variety of fossil vegetables, *Sigillaria, Lepidodendron* and fronds of ferns. Amongst these also occur large masses of fossilised remains of vegetation of irregular forms, generally flattened, and seldom exceeding 2 or 3 feet in length. These masses are invested with an irregular coat of carbonaceous and clayey matter, in which are inserted small fragments of stems, showing that it is not the true bark carbonised, but a confused assemblage of vegetable remains. They are evidently composed of portions of plants, of very different diameters and textures, compacted in a mass of decayed vegetable substances and broken up by crystallisation of calcareous spar, and present each, if I may so say, a whole magazine of species". (Witham 1833, pp. 7 and 39–42).

From Witham's description and figures (by the botanist Macgillivray) Dr Scott concluded that *Anabathra pulcherrima* was a lycopod with secondary growth in thickness, and no more precise identification than this is known to the writer.

The principal Berwickshire localities from which Witham collected fossil plants were Lennel Braes and Tweed Mill. A preliminary account of the Lennel Braes fossils was given in a paper in the *Philosophical Magazine* (Witham, 1830 (b) p. 16–21). On p. 16 of this paper he acknowledged the assistance of his "intelligent friend Mr Francis Forster", and then on p. 17 proceeded to say: "Lennel Braes being the most exposed to the swelling waters of the Tweed, these ancient fossils are to be obtained there in the greatest abundance." To illustrate this location of the fossils he gave a diagrammatic section of the exposure at Lennel Braes in a line N. fifteen degrees E., the rocks dipping in that direction eight degrees. A summary of this section is given below. From it, it appears that the fossil plants occurred in the lowest beds of shale near the water level of the Tweed which is marked on the diagram. Witham went on to state that:

> The highest stem I have been able to obtain is not much more than 4 feet, and the lowest part of it is about 6 feet in circumference. No two stems possess the woody appearances alike, some retaining it in the centre of the stem, others having such appearances distributed in various parts of the stem. Owing to the immense superincumbent mass, this part of the research is rendered both tedious and expensive.

Witham rightly concluded that the fossils were Carboniferous in age, though he claimed some of the rocks on Tweedside (eg. at Milne Graden) were in the New Red Sandstone. He also mentioned visiting Greenlaw, Polwarth, Langton Burn, St. Helens (2 miles south-east from Dunce), and Chirnside Bridge, and thought that many of the rocks at these places were "New Red Sandstone" A similar view was expressed by R.D. Thomson in a paper read before the Berwickshire Naturalists' Club, 21st December 1831 (Thomson 1832) but in that year Witham abandoned that view (Witham 1831 (a)).

Here follows the summary:

	ft	in
Sandstone	8	0
Shale with 2 inch bed of iron stone	1	6
Sandstone divided by shale	2	0
Shale	4	0
Sandstone	1	0
Shale with 3, 6 and 8 inch balls of ironstone and Siliceous matter	9	0
Shale	7	0
Sandstone	1	6
Sandstone in thin plates	2	0
Shale confused in its beds (resembling the sill of a coal seam containing the vegetable organic remains with their irregular layers of coaly matter)	8	0
TOTAL	44	0

Concerning the interpretation of the fossil trees at Lennel Braes he rightly regarded them as not being Vascular Cryptogams. This he emphasised as a remarkable fact since Adolphe Brongniart – "the father of palaeobotany" – had stressed the overwhelming superiority in numbers of the Vascular Cryptogams in the Carboniferous period, together with plants he called Monocotyledons – which were probably Cordaitae (an extinct group of fossil Gymnosperms). Witham regarded the fossil trees from Lennel Braes as Dicotyledons, in which group Gymnosperms were then included. Thus his discoveries at Lennel Braes supported his conclusions regarding "the Craigleith tree". This tree was discovered in 1826, and is now called *Pitus Withami*; it was removed from the quarry at Craigleith, near Edinburgh, and set up in the grounds of the Natural History Museum at South Kensington. It measured 47 feet in length and the wood was still 1 foot in diameter at the top (Scott, 1923, p. 254). Concerning this tree Witham wrote:

"We have in this striking and stupendous relic of ages long gone by, an additional proof amongst many others lately advanced that plants belonging to the Gymnospermous Phanerogamic class are much more abundant in these early sedimentary deposits than continental writers would lead us to think". (Witham, 1831 (c), p. 10). It was therefore, Witham's greatest achievement to demonstrate that fossil Gymnosperms were prevalent in early Carboniferous times.

Besides the locality at Lennel Braes, Witham also obtained similar fossil trees at Tweed Mill, of which he wrote, "Numerous beautiful remains of stems of similar plants are found at Tweed Mill, on the North bank of that river." (Witham, 1833, p. 7). Again: "The remains of ancient vegetation which are found at Tweed Mill, about a mile below Lennel Braes, resemble those at the latter locality in their external appearance, and it is probable that they belong to the same deposit, or even the same stratum." (Witham, 1833, p. 37). Witham proposed the name *Pitus antiqua* for the "Lennel Braes tree", which was also represented among the Tweed Mill stems, and was distinguished by medullary rays 3–5 cells wide. For a second species, with wider medullary rays (8–15 cells wide) he proposed the name *Pitus primaeva* (Witham, 1833, pp. 38–9).

These fossils were well known to other geologists, some of whom were members of the Berwickshire Naturalists' Club. Thus, Winch, in his paper on the geology of the banks of the Tweed, referred to Witham's description of the strata at Lennel Braes, and added; "The petrified trunks of trees are irregularly dispersed through the lower beds of shale, and are both of the Monocotyledons and Dicotyledons." (Winch, 1831 (e), p. 120).

In 1835 Dr R.D. Thomson discussed the rocks of Berwickshire and wrote, "That the Merse rocks are intimately connected with the Carboniferous group is obvious from the circumstance of our meeting with considerable remains of plants in the quarry at Whitsome, bearing a strong resemblance to *Calamites*." Again, "At Lennelhill, where fossil vegetables have been so unmercifully quarried by amateurs, as to leave scarce a vestige for the man of science, the only person to whom they could be of the slightest value, the limestone and shale containing microscopic shells are clearly members of the carboniferous group". (*H.B.N.C.*, Vol. I, pp. 85–6.).

In the year 1835, which saw the founding of the Geological Survey of Great Britain, David Milne published *A Geological Survey of Berwickshire*; an essay submitted to the Highland and Agricultural Society of Scotland, and for which he was awarded the first prize of fifty guineas. David Milne (1805–90) later assumed the name Milne Home after his marriage, and gave up law, thereafter devoting much of his time to his favourite pursuit of geology. He was admitted to the Club in 1836 (*H.B.N.C.* Vol. I, p. 106) and became President in 1860–1 (*H.B.N.C.*, Vol. IV, pp. 219–60). His death is recorded in Vol. XIII, p. 4 and an obituary notice is given on pp. 407–9 of the same volume. In his book (p. 24) David Milne referred to Witham's fossils as follows:

> It is in these beds of blue shale or clay that the fragments of large coniferae were first discovered by Mr Witham of Lartington. The two localities from which he excavated the greatest number are Tweed Mill, in the parish of Coldstream, and Allanbank, in the parish of Edrom. They occur in pieces not exceeding 3 or 4 feet in length and 2 feet in diameter. The original form of the tree is seldom preserved. The fragment is usually flattened, and little of the woody fibre is left, nearly the whole being displaced by carbonate of lime, which forms a fine granular limestone. This limestone is so compact that it is difficult to break even with the hammer, but, when it is broken, pieces of carbonised matter, and of a honeycombed-looking substance, are detected, being the external bark and woody parts of the original tree. No entire tree has yet been discovered. What are generally found seem to be pieces of trunks, rounded at the ends as if they had been transported, and worn down by attrition; none of them is erect – they lie in the beds of clay parallel with the strata. Small branches have been occasionally found by the Author in immediate contact with these larger fragments, the shape and whole parts of which, though much impregnated with carbonate of lime are completely preserved . . . Beautiful sections have been made by Mr Nicol and Mr Sanderson of Edinburgh, of those Tweed Mill and Allanbank fossils, which exhibit, even to the naked eye, the reticulated structure, and the annual growths of the tree . . . Allanbank and Tweed Mill are not the only places where these interesting fossils have been found. The Author has noticed them on the south bank of the Tweed, a little way below Coldstream Bridge, as well as on the north bank of the river at Lennel Church.

In the year 1846 Robert Embleton in his Presidential Address to the Club said, "Some very fine fossil trees similar to those found at Lennel some years ago were inspected at the vicarage (Norham) . . . They were dug out of the sandstone cliff where the new and hideous bridge is erected" (*H.B.N.C.* Vol. II p. 168). These were again referred to by Dr Gilly, Vicar of Norham, on p. 182. "One large stone, within the iron railing of Norham Churchyard . . . is part of a fossil tree dug up from the northern bank of the Tweed in 1839–40, when they were constructing the road which leads from Ladykirk and Norham Bridge."

Dr George Johnston reported on this fossil as follows; "I have now ascertained that your fossil trees are identical in kind with those from Lennel Braes. They are beautifully figured in a work of the late Mr Witham, and about sixteen years ago created a great deal of interest, for they were amongst the first of the coniferous fossils that had been found in so old a formation." (*H.B.N.C.*, Vol, II, p. 182).

In 1853 Dr Johnston published his first volume of *The Natural History of the Eastern Borders*. To this was appended *The Fossil Flora of the Mountain Limestone Formation of the Eastern Borders*, by George Tate (pp. 289–317). On p. 297 Tate commenced his list of fossil plants with an account of Witham's "Coniferae" from which one or two points are worth quoting. His description of the Allanbank source agrees with that of Witham (1831 (a), p. 180) in that he said that the quarry was, "at Allanbank Mill, near the junction of the Whiteadder and Blackadder". In a footnote he added: "We have recently examined the strata at Lennel Braes and Tweed Mill, and found the Coniferous trees in masses of limestone associated with marine fossils – *Spirorbis, Carbonarius*, and undetermined species of *Orthoceras, Pleurotomaria* and *Avicula*, proving that these plants had been carried into the sea, and there fossilised."

On pp. 299–300 Tate gave an account of the genus *Sigillaria* and said that it occurred at Lennel Braes: "In our district, remains of the stems of *Sigillaria* are frequent in the sandstone connected with the coal; they are, however, generally decorticated and rarely exhibit the form of the scars, by which the particular species may be determined. One species we have been able distinctly to determine; *Sigillaria* organa (Sternb.). Locality Lennel Braes etc." In his account of Witham's fossil, *Anabathra pulcherrima*, he stated: "This plant is nearly related to *Sigillaria*."

In 1859 Tate gave a lucid account of the Lower Carboniferous strata in the Eastern Borders (*H.B.N.C.*, Vol, IV, p. 149). He designated the Calciferous Sandstone Series "The Tuedian Group", and wrote:

> In 1856 I applied this name to a series of beds, lying below the Mountain Limestone, which are largely developed on the Tweed. They consist of grey-greenish and lilac shales, sandstones, slaty sandstones, sometimes calcareous, thin beds of argillaceous limestone, and chert, and a few buff magnesium limestones. *Stigmaria ficoides, Lepidodendra*, coniferous trees and other plants occur in some parts of the group, but there are no workable beds of

coal ... In one bed on the Tweed, *Orthocerata* and *Pleurotomariae* – marine Mollusca – are associated with coniferous trees. The whole group is especially distinguished by the absence of *Brachiopods*, which are abundant in the overlying Mountain Limestone. It forms a marked transitional series, intercalated between the Mountain Limestone and the Old Red Sandstone. Generally freshwater and lacustrine conditions are indicated and when marine remains do occur, they are accompanied with plants which appear to have been swept into a shallow estuary.

Further references to Witham's fossils in the *History of the Berwickshire Naturalists' Club* are found in Vols. V and VII. Thus in 1865 (vol. V, p. 187). Frederick J.W. Collingwood, Esq., President, referred, in his Presidential Address, to a meeting at Norham, when one party, led by Mr G. Tate, strolled up the Tweed:

The rocks examined belong to the Tuedian group, the lowermost division of the Carboniferous formation; teeth and scales of *Rhizodus hibberti* and of other ganoid fish were discovered, and the remains of *Lepidodendron* and other plants ... A little above Milne Graden – at Tweed Mill – are remains of Araucarian trees associated with species of *Orthoceras, Pleurotomaria* and other marine molluscs ... At Milne Graden a remarkable and magnificient specimen of a fossil tree, about 20 feet in length, found in an adjacent quarry, was shown by Mr D. Milne Home.

In 1873 (Vol. VII, p. 20) Dr Charles Stuart, in his Presidential Address, gave an account of a meeting at Chirnside held on 28th August 1873.

Crossing Allanton Bridge, we walked up the banks of the Blackadder, below Allanbank House. Here on the south side of the river we have a good section of the Tuedian strata, including sandstones and fossiliferous shales ... Mr Witham long ago found plenty of Coniferae etc. in these strata, in a sandstone quarry, in the immediate vicinity, at Allanbank Mill, now filled up.

"This bed," Mr Stevenson writes, "is just a little below the dark shales in which the fish, etc. remains occur, and is the same sandstone as is quarried at Langton, Puttonmill, Kimmerghame, Broomhouse, Eccles, Coldstream, etc. Its geological position is at the bottom of the Lower Carboniferous series."

From the above it would thus appear that the quarry from which Witham obtained the Allanbank fossils was situated at Allanbank Mill near the junction of the Blackadder and Whiteadder. It is therefore, not to be confused with the old quarry (still in existence though unused) near the ruined Blackadder Mansion on the south side of the river opposite Allanbank.

Although Witham's Quarry was filled in at the time Dr Stuart wrote (1873), it is of interest that the present writer has found petrified plants on the north bank of the river immediately below Allanbank Mill. In these petrifications are remains of *Stauropteris burntislandica* and a seed-like fructification to be described under the name *Genomosperma latens*, which has also been found on the Langton Burn, near Gavinton. After the formation of the Geological Survey in 1835, Berwickshire was partially surveyed by Sir Archibald Geikie, who published his results in 1863 in a very interesting memoir, entitled, *The Geology of Eastern Berwickshire* (Map 34). To this was appended a list of fossils, compiled by J. W. Salter, A.L.S., F.G.S. No mention is made, however, of any further examples of Witham's species, as these were found outside the area covered by the Memoir.

List of fossil plants compiled by J.W. Salter appended to Geikie's Geology of Eastern Berwickshire:

OLD RED SANDSTONE

(a) Lower

Linear grass-like plants, not leaves of *Sigillaria* from Ale Water, near Ayton.

(b) Upper

Roots from Banks of Whiteadder, near Cockburn Mill

Cyclopteris – From same locality – found by Mr Stevenson

CARBONIFEROUS

Calciferous Sandstones

Sagenaria Veltheimiana Goepp. – Banks of Whiteadder, 1 mile north-west of Chirnside.

Cyclopteris. From above locality and shore at Burnmouth. Coniferous leaves (allied to *Salisburia*) Hutton Mill, Banks of Whiteadder.

Sphenopteris – Cove, Cockburnspath

Lepidophyllum – Cove Cockburnspath

Cycadites Caledonicus nov. sp. – Cove, Cockburnspath, at the bottom of the Carboniferous series.

(A description of this fossil is then given, but will not be quoted, as it is now considered to be part of an animal, viz. an Arthropod – probably a Eurypterid, and is under investigation by Dr C. D. Waterston of the Royal Scottish Museum).

As mentioned above, Geikie's survey of Eastern Berwickshire did not include the Coldstream area, but this was covered by the survey of 1895, entitled "The Geology of part of Northumberland including the country between Wooler and Coldstream," by W. Gunn and C. T. Clough.

On p. 24 the authors gave an account of the strata exposed on the banks of the Tweed, near Coldstream, which is worth quoting:

> It is probable that the beds exposed by the Tweed side, below Coldstream Bridge are higher in the series than any of these [referring to rocks higher up the Tweed, near Carham], and we get, apparently a continuous ascending section from the bend of the river above the bridge, going down the stream, for nearly a mile, the beds dipping pretty steadily between E.-N.E. and N.E. from 5 degrees to 7 degrees. They consist of grey and dark shales with thin-bedded greyish and yellowish sandstones, some mudstones, clays, cement and limestone bands. As we go further east (downstream) and ascend in the series to where sandstones abound, we find numerous plants. Lennel Braes, where Witham obtained many of the species described in Fossil Vegetables is 2 miles from Coldstream Bridge, but on the Scottish side of the river, and, according to Witham, the north-east dip prevails all this distance. At all events the north-east dip continues pretty constant on the English side as far as Caller Heugh Bank, about a mile from the mouth of the Till.

In the list of fossils, published on p. 86 as an appendix Gunn and Clough state that, "Plants were collected by the survey at several places along the banks of the Tweed. Stems showing traces of structure, but too imperfect for determination, from 100 yards below the junction of the River Till, and specimens of *Araucarioxylon* (ie. *Pitus*) as determined by Mr Kidston from the Tweed three quarters of a mile below Coldstream Bridge, and also from the same ¼ mile above the junction of the Till". In a list given on p. 88 and supplied by Mr B. N. Peach, F.R.S., *Alcicornopteris convolutus* Kidston is recorded from Lennel Braes.

In 1932 the Geological Survey published *The Geology of the Cheviot Hills*, by R.G. Carruthers and others. This was based on the work of

C. T. Clough and W. Gunn, but as far as Witham's species of fossil plants are concerned, the authors have nothing fresh to add. Mention is made, however, of a fossiliferous plant bed in the Willow Burn, south of West Learmouth (Northumberland) containing *Adiantes antiquus* and *Lepidodendron nathorsti* (see p. 52).

Since the beginning of the present century the genus *Pitus* has been subjected to further investigation by Dr D. H. Scott and Professor W. T. Gordon. In 1902 Dr Scott published his paper Primary Structure of Certain Palaeozoic Stems with the *Dadoxylon* Type of Wood (Scott 1902). In this paper he also described another Berwickshire fossil stem *Calamopitys Beinertiana*, which was collected at Norham Bridge by Dr Kidston in September 1900, and which subsequently was referred to the genus *Eristophyton* by Zalessky, so that it is now known as *Eristophyton Beinertianum* Zalessky (Seward, 1917 p, 199).

Scott's reinvestigation of *Pitys antiqua* was based largely on a stem from Dr Kidston's collection found by Mr B. N. Peach, F.R.S. at Lennel Braes in 1883 and another specimen in Dr Kidston's collection also from the same locality. The chief interest of these specimens is the presence of many mesarch primary xylem strands around the pith. In his general summary and conclusions Scott wrote:

> The *Pitys* stems appear to have belonged to tall branching trees. We know of no Cycads or Cycadofilices with a similar habit nor is there any evidence that the Coniferae existed at so early a period. The only known family to which these trees could be referred is that of the Cordaiteae, leaves of which have been found at a similar horizon. The species of *Pitus* differ from stems, traced with certainty to true Cordaiteae, in the broad medullary rays, the non-discoid pith (for the slight approach to discoid structure which they exhibit is of doubtful value), and in the presence of the primary xylem-strands. On the whole I am disposed to regard the genus *Pitus* as a primitive member of the Cordaitean family retaining some of the characters of an earlier stock. The mesarch xylem-strands, in spite of their reduced size, and the peculiarities of their arrangement, are evidently comparable to those of *Lyginodendron* or *Calamopitys*. Thus the *Pitus* trees appear to afford a new link, so far as stem-structure is concerned, between the Cycadofilices of the family Lyginodendreae and the true Cordaiteae. (Scott, 1902, p. 362).

In 1935 Professor W. T. Gordon published his comprehensive paper *The Genus Pitys Witham, Emen*. Gordon not only reviewed Witham's species, but added a new species to the genus which he named *Pitys dayi* from ash below Hanging Rocks, Weak Law, Gullane in East Lothian, the geological horizon being in the Lower Carboniferous Oil Shale Series. This species has the cortex, buds and leaves preserved. The leaves measure 4 mm. × 6mm. × 50 mm. and resemble in transverse section the petioles of *Lyginodendron* so that Gordon interpreted them as phyllodes.

His general diagnosis of the genus reads as follows:

Stems arborescent and of large size. Pith large, continuous, parenchymatous and of several types of element. Primary xylem a reticulum of marginal and sunk mesarch strands, not in contact with secondary wood. Tracheides spiral to scalariform. Secondary xylem well developed; tracheides with multiseriate bordered pits; pits adpressed or free; pore slits elongate, opposed slits (on same pore) crossed, and oblique to length of tracheide. Primary medullary rays wide and high near pith, becoming narrower outwards; secondary rays vary according to species. Ratio of medullary ray to wood in tangential section relatively great and so wood relatively soft. Bark narrow compared with diameter of stem; proportions not dissimilar to those in recent conifers. Sclerotic nests in cortex and periderm formation developed. Leaf-traces are branches from primary reticulum; strands single at first, but divide in cortex of stem; traces ruptured with growth in girth, and inner ends occluded in xylem of stem. Leaves needle-like so far as known, though there may be needle-leafed and broad-leafed types, as in the living araucarian forms. Roots diarch to polyarch; xylem elements similar to those of stems. Fructifications unknown. (Gordon, 1935, p. 301).

In his concluding summary Gordon agreed with Scott that *Pitus* was probably descended from a *Lyginopteris* type of plant, but he differed from Scott in considering that the genus was less closely allied to the *Cordaitales* than had been supposed and was probably on the line of descent of the *Araucarineae* among the *Coniferae* (Gordon, 1935, p. 307).

In concluding this account of Henry Witham's work on Berwickshire fossil plants it may be of interest to add that specimens of *Pitus* from Lennel Braes are present in the collections of the Royal Scottish Museum Edinburgh, and in the Hunterian Museum at Glasgow University. Some of these bear such a close resemblance to Witham's largest specimen figured in his work that it is highly probable that the portions were cut from the same specimen.

Finally, mention may be made briefly to the fact that a letter from

Witham to Matthew Culley, F.G.S., of Coupland Castle, Northumberland, is printed in the Club's History, Vol. XXI (1911) p. 288.

REFERENCES

(arranged in chronological order)

1816 Winch, N. J., "Observations on the Geology of Northumberland and Durham". *Trans. Geol. Soc.*, ser. 1, Vol. IV, p.1.

1830 (a) Witham, Henry T. M., "Vegetation of the First Period of an Ancient World". *Phil. Mag.*, ser. 2, vol. VII, pp. 16–21.

1830 (b) Witham, Henry T. M., "On the Vegetable Fossils found at Lennel Braes near Coldstream, upon the Banks of the River Tweed in Berwickshire". *Phil. Mag.*, ser. 2 vol. VIII, pp. 16–21.

1831 (a) Witham, Henry T. M., "On the Red Sandstones of Berwickshire, particularly those at the Mouth of the Tweed". *Trans. Nat. Hist. Soc. of Northumberland, Durham, and Newcastle-upon Tyne.* Vol. I, p. 172.

1831 (b) Witham Henry T. M., *Observations on Fossil Vegetables, accompanied by Representations of their Internal Structure as seen through the Microscope.* (Edinburgh and London)

1831 (c) Witham, Henry T. M., "Description of a Fossil Tree discovered in the Quarry of Craigleith", *Trans. Nat. Hist. Soc. Newcastle.* Vol. 1, p. 294.

1831 (d) Witham, Henry T. M. Review of "Observations on Fossil Vegetables", Ed. *Jour. of of Science*, Vol. V (new ser). pp. 183–9.

1831 (e) Winch, N. J., "Remarks on the Geology of the Banks of the Tweed." *Trans. Nat. Hist. Soc. Northumberland*, etc. Vol. 1; and also *Phil. Mag.* (new ser.) Vol. IX.

1832 Thomson, Robert Dundas, "Contributions to the Geology of Berwickshire" (read before the B.N.C., 21 December 1831), *Mag. of Nat. Hist.*, Vol. V, p. 637.

1831–33 Lindley, J. and Hutton, W., *Fossil Flora of Great Britain*, Vol. 1 (Vol. 2 1833–5; Vol. 3 1837).

1833 Witham, Henry T. M, *The Internal Structure of Fossil Vegetables found in the Carboniferous and Oolitic Deposits of Great Britain, described and illustrated.* (Edinburgh and London).

1834. Nicol, William, "Observations on the Structure of Recent and Fossil Coniferae", *Edin. New Philos. Journ.*, Vol. XVI, pp. 137–58.

1835 (a) Milne, David, "A Geological Survey of Berwickshire", *Prize Essays and Trans. Highland and Agric. Soc. of Scotland*, Ser. ii Vol. XI, (new ser. Vol. V): pp. 171–253.

1835(b) Thomson, R. D., "Observation on the Strata of Berwickshire and North Durham", *H.B.N.C.* Vol, 1, p85.

1845 Stevenson, William. "On the Geology of Cockburnlaw and the adjoining District of Berwickshire", *Trans. Roy. Soc. Edin.* Vol. XVI, pt. I, pp. 33–46.

1846 Embleton, Robert, "Presidential Address", *H.B.N.C.* Vol. II, p. 168.

1850 Stevenson, William, "On a gap in the Greywacke Formation of the

Eastern Lammermuirs filled by Old Red Sandstone Conglomerate", *Quart. Journ. Geol. Soc.* Vol. VI, p. 418.

1853 Tate, George, "Fossil Flora of the Mountain Limestone Formation", in *The Natural History of the Eastern Borders*. Vol. I, by Dr George Johnston, pp. 291–317.

1859 Tate, George, "Fauna of the Mountain Limestone Formation on the Berwickshire Coast, with a Preliminary Notice of the Succession of the Strata on the Eastern Borders", *H.B.N.C.* Vol. IV, pp. 149–53, and p. 168.

1863 (a) Geikie, Sir Archibald, "The Geology of Eastern Berwickshire", *Mem. Geol. Surv. Gt. Britain.*

1863 (b) Tate George, *H.B.N.C.* Vol. V. pp. 73, 85.

1865 Collingwood, Frederick, J. W., "Presidential Address", *H.B.N.C.* Vol. V, p. 187.

1869 Miller, Hugh, *Testimony of the Rocks*, p. 411.

1873 Stuart, Dr Charles "Presidential Address" *H.B.N.C.* Vol. VII, p. 20.

1895 Gunn, W. and Clough, C. T. "The Geology of Part of Northumberland, including the Country between Wooler and Coldstream", *Mem. Geol. Surv. Gt. Britain*, p. 24 and 86.

1902 Scott, D. H., "Primary Structure of Certain Palaeozoic Stems". *Trans. Roy. Soc. Edin.*, Vol. XL., p. 362.

1910 Clough, C. T. and others, "The Geology of East Lothian". *Mem. Geol. Surv. Great Britain.* p. 44.

1911 Scott, D. H., "Presidential Address to the Linnean Society," *Proc. Linn. Soc. Lond.*, 123rd Session, pp. 17–29.

1913 Scott, D. H. "William Crawford Williamson (1816–95)" in *Makers of British Botany* ed. by F.W. Oliver, pp. 243–60.

1917 Seward, A. C. *Fossil Plants*, Vol. III, p. 199.

1923 Scott, D. H., *Studies in Fossil Botany*, Vol. II, pp. 254–60.

1932 Carruthers, R. G., *The Geology of the Cheviot Hills*, p. 52.

1935 Gordon, W. T., "The Genus *Pitys*, Witham, Emend", *Trans. Roy. Soc. Edin.*, Vol. LVIII pp 279–311.

1948 (a) Pringle, J., *The South of Scotland*, p. 60.

1948 (b) MacGregor, A. G. and Eckford, R. J. A., "The Upper Old Red and Lower Carboniferous Sediments of Teviotdale and Tweedside, and the Stones of the Abbeys of the Scottish Borderland" *Trans. Edin. Geol. Soc.* Vol. XIV, pt. II. pp. 230–52.

APPENDIX

Copies of the correspondence of the late Dr Dukinfield Henry Scott, F.R.S. relating to his search for a portrait of Henry Witham of Lartington (published by permission of the late Professor John Walton, Glasgow University and Mrs V.G. Wiltshire, daughter of Dr Scott).

Letter 1. From Professor John Wesley Judd C.B., F.R.S., formerly Professor of Geology at the School of Mines and the Royal College of Science, London.

30 Cumberland Road,
Kew.

17th February 1911.

My Dear Scott,

There is a short obituary notice of Binney in Etheridge's address to the G.S. (Q.J.G.S. Vol. XXXVIII, 1822 Proc., p 58). I cannot find one of Witham, though I fancy I have seen one somewhere, or an account of his connection with William Nicol, who made the sections for him. I had to look up the matter in connection with three obituaries of Sorby I was called upon to write. (I regret that my separate copies of these are all exhausted except the one I send).

There was a controversy carried on in the *Edinb. Phil. Journ.* by Alexander Bryson and others concerning the priority of William Nicol's methods and I fancy that Witham is mentioned as a well-known person.

All the Scotchmen, including Geikie, give Nicol the credit of inventing transparent sections of stony materials and say Sorby got his idea from Nicol through "Sandy" Bryson who had an optician's shop in Edinburgh. But Sorby, in a lecture he gave at Sheffield, "Fifty years of Scientific research" clearly states that it was to Williamson and not to Nicol, that he was indebted. I have stated this in the obituary I send (that for the Mineralogical Society) but in that for the *Geological Magazine* (May 1908pp 195–6) of which I can find no separate copy, I regret to say, the matter is stated more fully. Many years ago I was in "Sandy" Bryson's shop, and he showed me Nicol's preparations which were left to him (they are now in the Brit. Mus. Min. Dept.) – but I am sorry to say I did not talk to him about Henry Witham.

I have heard that Nicol acted as assistant to a popular lecturer of that day (Richardson I think was his name) and that Witham, who was I think a gentleman of private means was directed to Nicol (who had made the celebrated prism, without understanding the theory of its action) as one likely to be able to make sections for him.

Have you looked in the National Dictionary of Biography? Last evening I was at the Athenium, and thought I would see what had been done about W. C. W. I found the notice a long and interesting one, with some personal details I had not seen before. It is written by Marcus Hartog, who states that he acted as demonstrator to Williamson, and he gives among his authorities our friend P. J. Hartog (now of London University), his brother. I do not know if Witham and Binney find a place in the Dictionary, but will look.

As showing how little Williamson was appreciated I may mention that when I proposed him for the Wollaston Medal in 1890, I met with much opposition in the Council of the G.S. – John Evans exclaimed, "Why he isn't even a fellow of the G.S.!". But I turned the tables, by reminding them that the first recipient of the W.M. (William Smith) was not a fellow and at last carried my point. The old man was very delighted and asked me to receive the medal for him. I read a letter (see *Q.J.G.S.*, 1890 Proc. pp. 34, 35) which he wrote to me for the purpose. Kind regards

Yours faithfully
John W. Judd.

Letter 2. From Sir Albert Charles Seward, Professor of Botany, in the University of Cambridge.

Westfield,
Huntingdon Road,
Cambridge

17th February 1911.

Dear Scott,

It is true that I once visited the house of Witham at Lartington and saw in a cabinet a few of the less interesting of the "Fossil Vegetables". Monsignor Witham of Lartington, who acted as guide, was a Roman Catholic Priest. Palaeozoic in age – very garrulous – he took no interest in his kinsman's work but preferred other topics of conversation. As I keep no diary I have no notes of the visit or you should have them.

Lartington is an old-world place, in keeping with Witham, as he is regarded by the present generation. These pioneers were fine fellows, and it was with the feelings of a pious pilgrim that I penetrated to the home of the author of *The Internal Structure of Fossil Vegetables*. "On accumule erreures sur erreures, " but as an offset against this tendency, perhaps I may be remembered as a student of fossil plants who once made a journey to the Palaeobotanical MECCA.

On 25th March I lecture at Bedales

Yours ever
A. C. Seward

Letter 3. From Captain F.M. Norman to Professor Isaac Bayley Balfour.

CHEVIOT HOUSE
Berwick Upon Tweed.

14th March 1911

Dear Professor Balfour,

You will gather from the encl. that Henry Witham of Lartington in 1831 pub. on Fossils.

His son, Thomas Witham, b. 1807, was Roman Catholic Priest in Berwick 1843–7, but I have no evidence to show whether he or his father wrote upon Tweed Mill Fossils.

He died in 1907 according to my correspondent Philip Witham – if his dates are correct.

There may be a portrait of him at Lartington; and Philip has an engraving of him, which you will observe, he does not offer to lend. Do you wish to go on with the quest? If so, it will be necessary to communicate somehow with Mr Silvertop, and after all how are you going to proceed? For you can hardly expect that the owners of the portrait or of the engraving will let it go out of their hands to strangers. I shall be glad to hear from you as to whether you consider it worth your while to go on.

As for the portrait which Philip saw at Lartington; what became of that when Lartington was sold? I enclose 2 letters, both of which you can send back please.

Yours sincerely
F. M. Norman

Postscript; Can you find out whether the name of the Witham who wrote on Tweed Mill Fossils was Henry or Thomas?

I have had to ferret out my correspondents with difficulty: hence the delay.

Letter 4. From Rev I. Stark, late R.C. Priest at Berwick-upon-Tweed, to Captain Norman. (Copy).

St. Mary's
Hexam,
Northumberland
8th March 1911

Dear Captain Norman,

I was surprised but very pleased to receive a letter from you and thus renew an old acquaintance.

I do not think I ever heard of the picture of Rev. Thomas Witham to which you refer. The Rev. T. Witham inherited the Lartington estate and lived at the Hall for many years. On his death the property passed to his great nephew Francis Silvertop of Minster Acres (Riding Mill) near Hexham. I understand that Lartington has been sold quite recently, but I do not know the particulars. Mr Silvertop is not at present living at Minsteracres, which I believe is unlet. There is a Mr Philip Witham of the firm of Witham, Roskell etc. Solicitors, London who I think is a member of the Witham family. I expect you would get from him all the information that is available. From Who's Who I see his address is Whitmoor House, Sutton Park, Guildford. I am sorry I cannot supply you with more definite information.

With kind regards and best wishes

Yours very sincerely, I. Stark
(Late R.C. Priest at Berwick)*

---

* The letter is a copy made by Professor Balfour from the original supplied by Captain Norman. Professor Balfour must have forwarded this copy along with Captain Norman's letter to Dr Scott.

Letter 5. From Mr Philip Witham to Captain Norman (Copy).

> Whitmoor House,
> Sutton Park,
> Guildford.
>
> 13 March 1911

Dear Sir,

In answer to your enquiry, Henry Witham of Lartington, in 1831, published *Observations on Fossil Vegetables* (Wm. Blackwood, Edinburgh, and Cadell, Strand). There is a portrait of him at Croxdale, the seat of Gerald Salvin. Henry Witham had a son, Thomas, born 1807, who was a Catholic Priest, and who on the death of his 3 brothers, came into the Lartington property and died in 1907 so that he must have been 100 (note by Captain Norman). I did not know that he had ever published anything himself but he may have done so. I think I remember seeing a portrait of him at Lartington the last time I was there, and I have an engraving of him, which must have been made from a painting.

Very pleased to give you any further information in my power.

> Yours truly
> Philip Witham

Letter 6. From Professor Isaac Bayley Balfour to Dr Scott.

> Regius Keeper,
> Royal Botanic Garden,
> Edinburgh.
>
> 15th March 1911.

My Dear Scott,

The enclosed letter from Commander Norman, R.N. and the copies of the two letters he sent to me and which I also enclose, will put you on the track of getting hold of what you want – at least I hope so.

You will read that there is a portrait of Henry Witham at Croxdale, and you will be able to get into personal communication with Mr Philip Witham who supplies that information, and achieve your object of getting a photograph or copy of it.

In replying to Commander Norman, who you see desires to know whether the subject is to be pursued further, I told him that I was sending on to you all the information, and that I should do nothing until I heard from you; but I think that in the circumstances it would be better were you yourself to write to Commander Norman. He is a keen botanist, and a keen antiquary, and he may be able to help you further in any quest in regard to Witham and his fossils.

Do not trouble to return to me Commander Norman's letter, but let me know what you do in the matter. I shall be interested to know if you find the portrait.

Yours ever,
Isaac Bayley Balfour

Letter 7. From Mr Philip Witham to Dr Scott.

Whitmoor House,
Sutton Park,
Guildford,

21st March 1911

Dear Sir,

The Henry Witham of Lartington Hall, who wrote the book on Berwickshire Fossils, was the second son of John Silvertop of Minster Acres, Northumberland, by Catherine, daughter of Sir Henry Lawson of Brough, Yorkshire. On the elder brother coming into the Brough Estates, the Lartington property shifted to Henry Silvertop.

Henry was born in 1779 and married Eliza, daughter of Thomas Witham of Headlam, and niece and co-heiress of William Witham, of Cliffe, Yorkshire, and by Sign Manual, he took the name and Arms of Witham.

The portrait of him and his wife are at Croxdale, as his second son, William, married Anna Maria Salvin, of Croxdale.

He was quite a nice looking man, and if you wish it, I will try and get a photograph of the portrait for you.

I regret that the above account will be of little help to you in reference to him as an Author on Fossil Botany.

Yours faithfully
Philip Witham

Dukinfield H. Scott, Esq.,
East Oakley House,
Basingstoke.

Letter 8. From Mr Philip Witham to Dr Scott.

> 1 Gray's Inn Square,
> London W.C.,
>
> 28 March 1911

Dear Sir,

I am sorry I haven't been able to answer your letter of the 22nd, as I have been very much engaged. Henry Witham died on the 28th November 1844. I will try and get you a photograph of the picture. Henry Witham, like all our family, was a Catholic.
  Believe me,

> Yours faithfully,
> Philip Witham

Letter 9. From Mr G.S. Boulger to Dr Scott.

> 11 Onslow Road,
> Richmond,
> Surrey.
>
> 2nd April 1911

Dear Dr Scott,

It is true that Britten and I have discussed the possibility of a continuation of Putteney's Sketches when we shall have got the new edition of our Index out of the way; but this is all *in nubibus*; so that, beyond some lives of deceased botanists, Masters, Percival Wright, C. B. Clark, A. W. Bennett, etc. which I am preparing for the Dictionary of National Biography, I am not likely to publish anything on modern history beyond the bare details of the Index for some years to come. You may be interested, in the following which I find among my MS notes: Witham Henry T.M. (1779?–1845) d. Lartington Hall, Yorkshire, 28 Nov. 1845; bur. Romald Kirk. First Catholic High Sheriff of Durham (Gateshead Observer, 7 December 1845). I am glad to gather from your letter that your lecture, and the other admirable component parts of the series, I hope, are to be preserved in a permanent form.

> Yours sincerely
> G. S. Boulger

Note The date quoted above (for Witham's death) was wrong, and Dr Scott must have written to Mr Boulger for a check to be made.

Letter 10. From Mr G.S. Boulger to Dr Scott.

> 11 Onslow Road,
> Richmond,
> Surrey.
>
> 2nd April 1911

Dear Dr Scott,

Your dates were correct, mine incorrect. I turned up the *Times* for Monday, December 2nd, 1844 and found: "On Thursday last, at his seat, Lartington Hall, Yorkshire. Henry Witham, Esq., High Sheriff of the County of Durham, aged 65".

I am not sure as to the source of my error; but suspect Joseph Gillow's *Bibl. Dict. of English Catholics.*

> Yrs. sincerely
> G.S. Boulger

(d) The Fossil Plants of Berwickshire – a review of past work.

Part 2. Work done mainly in the present century. From the *History of the Berwickshire Naturalists' Club*, Vol. XXXV. Part I, 1959. pp. 26–47.

In Part I of this paper it was seen that the principal discoveries of fossil plants in Berwickshire during the nineteenth century largely resulted from the work of Henry Witham. Similarly the discoveries of the present century stem largely from the work of one man, Robert Kidston, LL.D., D. Sc., F.R.S. (1852–1924). Both Witham and Kidston prosecuted their work as amateurs, and both left their mark on palaeobotanical science to a degree far in excess of the purely local interest of their work considered here.

An account of the life and work of Dr Kidston was written by Dr R. Crookall and published in 1938 by the Geological Survey (Crookall, 1938). The details quoted below are based on this account. I am also indebted to Dr Crookall for permission to reproduce the photograph of Dr Kidston which was loaned to me by Professor John Walton of Glasgow University. This photograph showing Dr Kidston at work in Bristol University, was taken by Dr Crookall only a few weeks before Dr Kidston's death.

Robert Kidston was born in 1852 at Bishopston House, Renfrewshire, but while he was still at an early age his parents removed to Stirling, which henceforth became his home town. For a time he was employed by the Glasgow Savings Bank but after 1878 he pursued his botanical researches full time, being enabled to do so by private means. His interest in fossil plants was probably aroused by attending lectures given by Professor W. C. Williamson in Glasgow, and he published his first scientific paper in 1880 at the age of 28. About that time he began to fulfil the task of honorary palaeobotanist to the Geological Survey. He also acted as joint Secretary of the Stirling Natural History and Archaeological Society from its foundation in 1878.

In a letter written to me from Egypt by the late Prof. F. W. Oliver of Cairo University and dated 24.2.1945, he says,

> Where you are now living must be classic ground in palaeobotany, Black and Whiteadder; Lennel etc. favourite hunting grounds of Kidston's. In 1881, when I hardly knew there was such a realm as fossil botany, I remember spending a night at Norham, and next day walking to Coldstream, Kelso and Melrose. Kidston once told me he had picked up quite useful specimens from the broken-up road material and from walls in that district. Though I have visited the Northumbrian and Scottish coastline as far as the Firth of Forth, I know best West Central Northumberland just South of Otterburn, where my forebears farmed up to about 1760.

Dr Kidston's first contact with Berwickshire fossil plants appears to have been made through Mr James Bennie, with whom he published a joint paper on Scottish Carboniferous spores (Bennie and Kidston 1886). This must rank as one of the earliest papers on a subject which has become of increasing importance and which has now a large and complex literature of its own. In this paper the first locality referred to on p. 93 is the shore "half a mile east of Cove Harbour and one and a half miles N.E. of Cockburnspath . . ." The spores described were found in the basement beds of the Calciferous Sandstones "in sandy fakes beneath a hard sandstone in which *Stigmariae, Lepidodendra* and *Calamite*-like plants in fragments are abundant. In the spore bed scorpion remains are frequent, and in the plant bed the original of the Eurypterid *Glyptoscorpius (Cycadites) Caledonicus* was found" (cf. Salter's list appended to Geikie's *Geology of Eastern Berwickshire*). Bennie and Kidston went on to say: "It is noteworthy that in the Upper Old Red Sandstone, which occurs only a few feet below, few if any plants are preserved, yet here all at once spores are found in the sandy fakes in myriads, proving the existence of an abundant vegetation little younger in age than that of the underlying O.R.S."

The spores discovered were named *Lagenicula* I and described as being "in a fine state of preservation".

The plant bed referred to above by Bennie and Kidston is still exposed in the little bay at Horse Roads, N.W. of Pease Bay, and is very near the base of the Carboniferous system. In this plant bed I have found an assemblage of fragmentary fossil plants similar to those later discovered by Mr A. Macconochie and Dr Kidston on the Langton Burn, near Gavinton. Among these plants there occurs *Stenomyelon tuedianum* Kidston – the stem of a primitive Pteridosperm which so far has only been found in Berwickshire. Its original discovery goes back almost to the middle of last century, when the first specimen was found at Norham Bridge by Adam Matheson, a millwright and amateur geologist of Jedburgh. Not much is known of Matheson, though he is referred to by Alexander Jeffrey in the *History and Antiquities of Roxburghshire*, Vol. IV (Preface) where he is mentioned as having afforded "much information in regard to points of local interest in the geology of the district" (Jeffrey, 1864). He was also known personally to David Milne, who mentioned him in two footnotes in his *Geological Account of Roxburghshire* (Milne 1843, pp. 441 and 477). Milne comments on Matheson's "geological zeal" and describes how he attempted to trace the course of the Hawick volcanic dyke south of the Border: "Having intimated to me his intention of setting out on this voyage of discovery, and asked me for instructions, I sent him out a map, compass and other necessary instruments. He writes me, that he hired a horse at Jedburgh, and set out from Hindhope along the line which, at that place the dyke appeared to run in." Matheson apparently succeeded in tracing the dyke to within seven or eight miles of the sea.

Adam Matheson's discovery of *Stenomyelon* must have occurred sometime before 1859 as in that year some sections were presented to the museum at the Royal Botanic Gardens, Edinburgh. The first specimen to come into the hands of Dr Kidston was labelled "near Berwick" and was a gift of Dr B. N. Peach, F.R.S. who obtained it from his father, Mr C. W. Peach, A.L.S. This specimen was later ascertained to have come from Adam Matheson, who, Dr Kidston believed was also the author of an anonymous pamphlet describing some Fossil stems found at Norham Bridge.

In describing this fossil Dr Kidston wrote:

> The matrix containing Mr Matheson's fossil was an impure fine clay, containing a fair proportion of iron and one showing features which were possible of recognition in the field; but though a careful search for a similar bed was made in the neighbourhood

of Norham Bridge, no trace could be found *in situ*. Subsequently in 1901, we discovered some small blocks of the desired rock lying on the side of the road near the North end of the Norham Bridge. It was ascertained that the material came from a cutting made in the road while putting in a drain some time before; the surface of the road in the neighbourhood of the drain was therefore carefully examined, and in a small block which had been used for refilling the cutting the specimen was discovered which has enabled us to give a detailed description of the species.

(Kidston and Gwynne-Vaughan, 1912, p. 263; also Scott, 1923, Vol. II, p. 135; and Scott, 1924, p. 162).

Dr Kidston was aided in his search for Adam Matheson's fossil stem by Mr A. Macconochie of the Geological Survey. Arthur Macconachie (1850–1922) was born at Dailly in Ayrshire and worked as an assistant with the Geological Survey from 1869 to 1913. He was a fossil collector of great skill, with acute powers of observation in the field, and made several important discoveries, which are mentioned in an obituary notice written by John Horne (1924, pp. 395–397).

In 1900 Mr Macconochie discovered fossil plants at several localities in Berwickshire viz. at Lennel Braes, near Coldstream; at Willie's Hole, near Allanton; at the scaur near Edrom House and on the Langton Burn 400 yards N.N.E. of Gavinton. In the next year Dr Kidston accompanied him to these localities, and in addition they obtained specimens from the Ladykirk Burn, from the Blackadder above Allanton Bridge and from the Bell's Burn Scaur on the Whiteadder below Blanerne Bridge (Kidston 1901, 1902).

In his report to the Geological Survey for 1900 (Kidston 1901, p. 174), the following species were recorded from Berwickshire:

(i) From "the well known section of the Tweed at Lennel Braes"
Alcicornopteris convoluta Kidston
Sphenopteris (Diplotmema) patentissima Ett.
Lepidodendron sp.
Stigmaria ficoides Sternb.
Stigmaria ficoides var. *undulata* Goepp.

(ii) From the right bank of the Whiteadder a ¼ mile west of Edrom House
Marchantites n. sp.
Alcicornopteris convoluta Kidston
Sphenopteris sp.

Aphlebia sp.
Lepidodendron sp.

(iii) From the right bank of the Whiteadder ¾ mile below Allanton
Marchantites n. sp.
Aneimites sp. (later identified as *A. acadica* Dawson).
Sphenopteris elegans Brongt.
Alcicornopteris convoluta Kidston
Aphlebia sp.
Lepidodendron spitsbergense Nathorst
Lepidostrobus sp. (probably *L. allantonense* Chaloner)
Stigmaria ficoides Sternb.
Cardiocarpus bicaudatus Kidston (later re-named *Samaropsis bicaudata*)

Dr Kidston concluded this report by saying that other specimens still awaited examination, including some showing structure from Lennel Braes, Norham Bridge and Langton Burn.

The *Lepidodendron* which he identified as *L. spitsbergense* Nathorst, was shown to be clearly distinct from *L. veltheimianum* Sternb. Of the specimen *Aneimites* he said "Though small, it is the first evidence of this genus in British rocks." The fossil which he named *Marchantites* he regarded as "perhaps the most interesting fossil among the plant impressions collected . . . a genus which I believe has not been previously found in Carboniferous rocks."

In the report for 1901 (Kidston 1902) we read on p. 178 that, "In the Autumn (of 1901) Mr Kidston once more placed his valued services at the disposal of the Geological Survey, and accompanied by Mr Macconochie, made a search for rare fossil plants among the lowest Carboniferous rocks of the Border." Specimens were recorded from the following localities

(i) From the Whiteadder right bank scaur under Edrom Church ¼ mile west of Edrom House:
Sphenopteris elegans Brongt.
Lepidendron Veltheimianum Sternb.
Stigmaria ficoides Sternb.
Cardiocarpus bicaudatus Kidston

(ii) From Whiteadder right bank at Willie's Hole, one mile east of Allanton (locality iii of the first report).
Marchantites sp. distinct from that already noted

Lepidodendron Veltheimianum Sternb.
Stigmaria ficoides Sternb. var. *undulata* Gopp.

(iii) From road cutting at north end of Norham Bridge
A *Noeggerathia* – like pinnule was found in the upper portion of a cementstone band about twelve inches thick in a shale a few feet above the level of the road. "Some material showing plant structure was also collected." (This was probably *Stenomyelon*.)

(iv) From small stream ¼ mile N.E. of Ladykirk
Asterocalamites scrobiculatus Sch.

(v) From right bank of Blackadder ¼ mile above Allanton Bridge
Lepidodendron spitsbergense Nath.

(vi) From Whiteadder, Bell's Burn Scaur, near Chirnside (*Aneimites* sp.)

(vii) From Langton Burn, about 400 yards N.N.E. of Gavinton:
Lepidodendron spitsbergense Nath.

"In addition some small blocks of a shelly limestone were found in the shingle on the Whiteadder immediately below the right bank scaur under Edrom Church. The plant remains in these are not so well preserved as those in the Langton Burn material but the blocks contain much the same species. The Edrom material has, however, yielded a *Heterangium*, a genus not found previously in Berwickshire." This species was probably the one that Gordon included under the name of *Rhetinangium arberi* (Gordon) 1912, p. 814.

Dr Kidston also gave a list of species from Marshall Meadows Bay. In the summary of progress of the Geological Survey for 1902 (Kidston 1903) Dr Kidston published lists of fossil plants collected by Mr A. Macconachie in the neighbourhood of Cocksburnpath, and those which occur in Berwickshire are quoted below:

(i) From a quarter mile and 50 yards E.S.E. of entrance to Cove Harbour
Stigmaria ficoides var. *undulata* Göppert

(ii) From 90 yards S. of entrance to Cove Harbour, in shale and in ironstone band in the shale on horizon of Scremerston Series
Rhodea moravica Ett.

Cardiopteris polymorpha Gopp. (this was undoubtedly a *Cardiopteridium*, probably *C. nanum* f. *spetsbergense*, see Walton, 1941, p. 61)

(iii) From shore, a little below high water mark, 90 yards S. of entrance to Cove harbour. Horizon about 30 feet below the lowest of Cove Limestones
Sphenopteris dissecta Brongt. (*Diplotmema dissecta*).
Cardiopteris polymorpha var. *rotundifolia* Gopp. (*Cardiopteridium* sp.)
Asterocalamites scrobiculatus Schl.
Lepidophyllum lanceolatus L. & H.

(iv) From outcrop in slope above Cove Harbour in sandstone and red ironstone band; same band as (ii) above:
Calymmatotheca affine L. & H. (*Telangium affine*.)
Cardiopteris polymorpha var. *rotundifolia* Gopp. (*Cardiopteridium* sp.)
The isolated pinnules of this fern fill a band of red ironstone 2–3 inches thick.
Asterocalamites scrobiculatus Schl.
Lepidodendron Rhodeanum Sternb.
Lepidophyllum lanceolatum L. & H.
Lepidostrobus sp.

(v) From bay N.W. of Cove in shale resting on oil shale band. About 2 feet above Lowest Cove Limestone.
Calymmatotheca affine L. & H.

Arising out of these discoveries of Mr Macconochie and Dr. Kidston in Berwickshire, about the beginning of the century, a number of new species of fossil plants have been described by different workers at different times extending up to the present day.

In 1910 Count Solms-Laubach described and figured one of the Langton Burn fossils under the name of *Cladoxylon kidstoni* (Solms Laubach 1910). Accounts of this imperfect fossil stem are given by Seward (1917, p. 205) and Scott (1923, p. 160). The species is the only one of the genus known in Britain. According to Dr Scott, the specimen is part of a large stem containing an incomplete ring of steles. In each stele there is a narrow band of primary wood and a broad zone of secondary wood, in which the pitting is limited to the radial walls. Some of the pits are circular, as in Conifers, others transversely elongated. The narrow medullary rays are mostly uniseriate.

In view of the rarity of this fossil plant and our incomplete knowledge of it, new specimens would be of great interest. Professor Seward regarded the evidence for assigning it to the genus *Cladoxylon* as not convincing.

In 1911 P. Bertrand described an incomplete stem of a small fern under the name *Zygopteris kidstoni* (Bertrand, 1911 a and b). This was later figured by H.S. Holden in his account of the Upper Carboniferous fern *Ankyropteris corrugata* (Holden, 1930). Hirmer has re-named the plant *Protoclepsydropsis kidstoni* (Hirmer, 1927, p. 519). The specimen which Dr Kidston found in the Langton Burn material was incomplete, consisting of a decorticated stem with a solid stellate protostele without petioles. Further specimens of this primitive fern have been found recently by the writer at Hutton Mill, and near Allanbank.

In 1912 the specimen of *Stenomyelon tuedianum* discovered in 1901 at Norham Bridge was described in a joint paper by Dr Kidston and Professor D. T. Gwynne-Vaughan. It is clear that Dr Kidston envisaged a series of papers on the Carboniferous Flora of Berwickshire, since this was entitled Part I. Owing probably to the death of Professor Gwynne-Vaughan in 1915 and Dr Kidston's decision to investigate the silicified plants of the Rhynie chert bed in collaboration with Professor W. H. Lang, the series of papers on Berwickshire fossil plants was discontinued.

In their paper on *Stenomyelon tuedianum* Kidston and Gwynne-Vaughan declined to suggest any affinities for this fossil plant, beyond placing it in the Cycadofilices (Pteridospermae). They were also unable to describe the leaf adequately though they knew that it must have been large in size and they thought that "the lamina must have been of considerable thickness".

Other decorticated stems of *Stenomyelon* occurring in the Langton Burn material were later named *S. tripartitum* Kidston, but the species was not described by Dr Kidston. A brief description is given by Dr Scott (Scott 1923, pp. 141–143), and photographs were included by Krausel and Weyland in their account of *Aneurophyton germanicum*, with which *Stenomyelon* was compared (Krausel and Weyland, 1929, p. 323). Dr Scott considered that *Stenomyelon* came low down among the Pteridosperms, its nearest probable relationship being with the Calamopityeae.

In 1938 Dr Mary G. Calder reinvestigated the sections of *Stenomyelon tripartitum*, and came to the conclusion that it could not be considered distinct from *Stenomyelon tuedianum* (Calder, 1938, p. 310).

The only other species of *Stenomyelon* yet discovered is *S. muratum* Read, which comes from the New Albany Shale (late Devonian) of

North America. This species has a "mixed pith" and is, therefore, slightly more advanced than *S. tuedianum*. Read considers that *Stenomyelon* is probably in the lineage of the more primitive Calamopityeae rather than in a separate family (Read, 1936, p. 81).

The Calamopityeae are still very imperfectly known, as their foliage and fructifications have not yet been described. It is evident, therefore that the fossil plant which Adam Matheson discovered at Norham Bridge about 100 years ago is still far from being understood in its structure and relationships.

In the years 1923–25 Dr Kidston published his great work on *The Fossil Plants of the Carboniferous Rocks of Great Britain*. According to Dr Crookall, this monograph was to have been completed in about ten parts. The six parts published form the second volume of the Palaeontological Series of the Memoirs of the Geological Survey, and consist of 681 pages and 153 plates covering most of the Ferns and Pteridosperms. The Lycopodiales, Sphenophyllales and Equisetales remained undescribed.

On p. 18 Dr Kidston stresses the abundance of Pteridosperms (seed-ferns) in the Lower Carboniferous rocks, and adds, "Were it not for the fact that true ferns have been found as petrifactions in the Pettycur material and in Berwickshire, there would have been no absolute proof of their occurring in British Lower Carboniferous rocks." On p.19 he gives a list of the petrified fossil plants so far discovered in Berwickshire, all of them coming from the Cementstone Group. These are here quoted:

Fern *Zygopteris kidstoni* Bertrand
Pteridosperms
Rhetinangium cf. *Arberi* Gordon
Stenomyelon tuedianum Kidston
Lyginorachis papilio Kidston MS
Cladoxylon kidstoni Solms
Rachiopteris multifascicula Kidstoni MS. (*Kalymma tuediana* Calder)
Other Gymnosperms
Eristophyton (Calamopitys) Beinertiana Gopp.
Pitys antiqua Witham
Pitys primaeva Witham

Dr Kidston also gave descriptive accounts of several fossil plants recorded from Berwickshire and summarised the localities from which they were obtained. These records I have arranged below in alphabetical order:

(i) *Alcicornopteris convoluta* Kidston (pp. 418–420).

This is figured on Pl CVIII, fig. 2b, from shore ½ mile east of Cove Harbour and fig. 3, from right bank of Whiteadder ¼ mile below Allanton. Other localities given are:- Broomhouse Burn near Duns; North bank of Whiteadder between Edington Mill and Hutton Bridge (J.H. Craw): Lennel Braes scaur on S.E. side of Churchyard ¼ mile N.E. of Lennel Village; Kimmerghame Quarry, near Duns.

(ii) *Aneimites acadica* Dawson (p. 414).

This is figured on Pl. CX, figs 4–7. Of the specimen shown on fig. 4 Dr Kidston wrote, "This is the most perfect example I have seen. It was collected by the late T. Ovens, of Foulden, and after his death was given to me by his father to whom my thanks are due for the interesting specimen." Of its distribution Dr Kidston said it was, "very rare in Britain and restricted to the Cementstone Group of the Calciferous Sandstone Series." He cited three Berwickshire localities: Left bank of Crooked Burn about 50 yards below Foulden Newton; right bank of Whiteadder ¾ mile below Allanton; Bellsburn Scaur, near Chirnside.

(iii) cf. *Coseleya* sp. (pp. 371–2).

This is figured on Pl. LXXVI, fig. 7 and 7a. The fossil consists of a small specimen showing exannulate sporangia unassociated with foliage pinnules, and came from the left bank of the Crooked Burn, 50 yards below Foulden Newton.

(iv) *Diplotmema (Sphenopteris) dissectum* Brongt. (p. 248).

This is figured on Pl. LX, figs. 1–5, and text-fig. 15, p. 240. The species is only known from the oil-shale group and is recorded by Dr Kidston from about 30 feet below the lowest of the Cove Limestones on the shore, a little below high water mark 90 yards south of Cove Harbour, Cockburnspath. It is also recorded from a shale a few feet below the coals at Marshall Meadows, 2½ miles N.W. of Berwick-upon-Tweed.

(v) *Diplotmema (Sphenopteris) patentissima* Ett. (p. 253).

This is figured on Pl. LIV, fig. 6. and is recorded from Lennel Braes near Coldstream.

(vi) *Ootheca globosa* Kidston (pp. 371–2).

This is figured on Pl. LXXI, fig. 6 and 6a and consists of a fragment of a rachis showing globular exannulate sporangia at the apex. The specimen came from the left bank of the

Crooked Burn, 50 yards below Foulden Newton and was collected by T. M. Ovens.

(vii) *Sphenopteridium pachyrrachis* Gopp. (p. 168).

This is figured on Pl. XXIX, fig. 5 (from Long Craig Bay, 1 mile west of Dunbar). Kidston's Berwickshire record was from ¾ mile below Allanton on right bank of Whiteadder. It is possible that this was the fossil plant which Kidston had recorded previously as *Sphenopteris elegans*.

(viii) *Telangium affine* L. & H. (*Calymmatotheca affinis* Kidston) (p. 446).

This is figured on Pl. C; Pl. CII, fig. 1; Pl. CIV, Fig. 5; and text-figs. 41–43. It was also figured by Hugh Miller in his *Testimony of the Rocks* (frontispiece). This fossil plant is only recorded from the oil-shale group of the Calciferous Sandstone Series, where it is a very characteristic species. Dr. Kidston recorded it from two Berwickshire localities; sandstone and red ironstone band on horizon of Scremerston beds in outcrop in slope above Cove Harbour; shale resting on oil-shale band about two feet above lowest Cove Limestone in bay N.W. of Cove Shore west of harbour, Cove.

(ix) *Zeilleria moravica* Ett. (*Rhodea moravica* Ett.) (p. 441).

This is figured on Pl. LXII, figs. 3–5; and Pl. CXIII, fig. 4 (not Berwickshire specimens). The species occurs in both the Carboniferous Limestone Series and in the oil shale group of the Calciferous Sandstone Series. The only record from Berwickshire comes from a shale and ironstone band on the horizon of the Scremerston Coal Strata, 90 yards south of the entrance to Cove Harbour.

Although Kidston had recorded *Diplotmema adiantoides* Schlotheim (*Sphenopteris elegans* Brongt.) from the Whiteadder near Allanton and near Edrom in the Summaries of Progress (Kidston 1901 and 1902), there are no records stated in "The Fossil Plants of the Carboniferous Rocks of Great Britain". This would suggest that the fossil plants originally identified as *S. elegans* were something different; e.g. they may have been *Sphenopteris pachyrrachis* Gopp.

In 1927 Errol I. White of the British Museum (Nat. Hist.) published, "The Fish Fauna of the Cementstones of Foulden, Berwickshire". To this was appended a list of Lower Carboniferous Plants by W.N. Edwards (White 1927). The collection on which this paper was based was made by Thomas Middlemiss Ovens, an amateur geologist of Foulden, who died in 1912 at the early age of twenty, and who was mentioned by James H. Craw, former Secretary of the Club, in 1921 (*H.B.N.C.*, Vol. XXIV, p. 287).

A biographical sketch of T. M. Ovens was published in The Border Magazine for October 1927, together with a portrait showing Ovens at work on an exposure. I am indebted to Rev. David S. Leslie of Hutton for drawing my attention to this article and for the loan of a copy which was actually given to him by Martha Helen Ovens (the mother of T. M. Ovens) then resident at Mansfield, Foulden. According to Mr Leslie the father of T. M. Ovens was gardener to Major Wilkie at Foulden House. The writer of this biographical sketch acknowledged assistance in compiling his account from Mr. Maconnachie, Mr James Hewat Craw, Mr Robert Eckford of the Geological Survey and Rev. John Reid, Edinburgh, formerly of Foulden.

According to Mr Eckford,

> Thomas Middlemiss Ovens was born 6th June 1891, and died 30th March 1912. The dread malady that ultimately claimed him as a victim was the cause that drove him to fossil collecting when his hours at the bank were over. A word of praise is due to the late Mr Arthur Macconachie, of H.M. Geological Survey for the encouragement and help he gave to Mr Ovens. Mr Macconachie was quick in detecting the importance of Mr Ovens's finds and had the specimens submitted to the late Dr Traquair, at that time the authority on fossil fishes. Dr Traquair, reported that the fauna contained a number of species hitherto new to the British Isles. This was great news to Mr Ovens who applied himself with still greater zeal to unearthing these ancient life relics of far-off times. Unfortunately Dr Traquair died before the collection was completed. Then the War intervened, and the collection lay in Edinburgh, until 1921, when it was sent to the British Museum, London.

The Rev. John Reid reported on T. M. Ovens as follows:

> Mr Thomas Ovens was a quiet contemplative lad . . . preferring to go to nature for his pleasures and excitements. In the course of the river Whiteadder, with its haughs, cliffs and escarpments, he found a fertile source of interest and study . . . After leaving school he obtained employment in a bank at Coldstream, but the confinement proved too trying for a constitution which was never very robust, and he was advised to try some open-air occupation. This suited his natural inclination, and he entered with zeal into the study of geology . . . but the disease that has blighted so many promising lives had too firm a hold upon

Thomas Middlemiss Ovens at work on the exposure of shales in the Crooked Burn below Foulden Newton, Berwickshire. Reproduced from *The Border Magazine*, volume 32 (1927).

him . . . He was laid to rest in early manhood in the churchyard of that parish from which he had never long been severed.

From the Church *Life and Work* supplement, May 1912, the following excerpt is quoted. "It is pathetic to relate that two days after Mr Ovens died, a letter addressed to him was received from Mr Macconachie, . . . in which he offered him when he attained the age of 21, the Madam Pidgeon Fund of £30 yearly to help him to prosecute his geological researches."

From an article in the Berwickshire Advertiser we learn that "All Ovens's fossils were collected in the short space of two years 1910–1912. The fishes alone included four genera new to science and six new species. Up to November 1924 nearly half the specimens had been identified and as it was felt that it would be a pity to have the collection divided, the British Museum authorities approached Mr and Mrs Ovens, Foulden, and asked to be allowed to have the whole collection. Very generously, Mr and Mrs Ovens consented to do so."

From this article we also learn that Ovens had been employed at the British Linen Bank in Coldstream. Further, It was Mr John Bishop, Berwick, who first interested Ovens in geology, and with Geikie's

Geological Survey of Scotland borrowed from the village library, he had the best text-book that was then to be had for the vicinity of his own home at Foulden, to which area he very naturally turned for specimens.

When Geikie surveyed the area in 1864, fossil scales were found (at the Crooked Burn), and friends of Ovens still recall the delight of the young geologist when he got his first fossil specimen – a fossil scale. This spurred him on to other finds, and Nature was remarkably generous to him in revealing her secrets which had been buried for countless years.

In the opening paragraph of his paper Dr Errol White says: "So barren of fossil remains is the Cementstone Group of the Scottish Lower Carboniferous Rocks that any addition to our knowledge of the fauna and flora of the period is especially welcome.

The collection to be described below contained nearly 150 specimens and includes Plants, Lamellibranchs, Annelids, Arthropods and Fishes. All the specimens were obtained from sections exposed in the Crooked Burn, fifty yards below Newton Farm, in the parish of Foulden.

The beds in which the remains were found belong to a horizon quite near the base of the Cementstones and consequently the fauna is one of the earliest known from the Lower Carboniferous Rocks.

The lithology of the beds is somewhat inconstant, in a manner typical of these shallow water deposits; all are argillaceous and highly charged with lime. The rock in the majority of cases is a fine-grained, somewhat sandy shale and contains a fair sprinkling of mica. In a few instances the sandy element is coarse and predominates, while in others it is wanting, and the rock is a very fine-grained, horny mudstone, with conchoidal fracture. The series is therefore, typically estuarine in character.

This fine collection owes its existence to the zeal of the late Thomas Middlemiss Ovens, an enthusiastic young local geologist, and the state of the specimens is a tribute to his careful and skilled collecting. It is greatly to be deplored that Mr Ovens's activities have been cut short by his untimely death at the age of nineteen.

Mr and Mrs John Ovens, of Foulden, have generously presented their son's collection to the British Museum. The majority of the plants, however, had been sent to the late Dr Kidston during the lifetime of the collector, and are now in the Jermyn Street Museum. They are partly described in the *Memoirs of The Geological Survey of Great Britain* (Palaeontology, vol. II, 1923–26) and a note on the specimens in the British Museum is here appended by Mr W.N. Edwards.

In this appendix Mr W.N. Edwards wrote:

The plant remains in the Thomas Ovens Collection from Lower Carboniferous (Tuedian) rocks near Foulden are of considerable interest, though few species are represented. Some have already been described by Kidston and are in the Jermyn Street Museum of Practical Geology. In addition to those mentioned below, there are branched fragments of a Pteridosperm rachis, impressions of larger stems and some other obscure specimens. Numerous examples of *Spirorbis* occur on the plant remains.

Edwards then gave a list of the species; from the list certain points are quoted below:

(i) *Aneimites acadica* Dawson.

This is "the commonest plant at Foulden". (B.M. Geol. Dept. V 16860–64).

(ii) *Sphenopteris (Telangium) affinis* L. & H.

"This Pteridosperm frond (V16865) has previously been recorded only from the oil-shale group where it is abundant".

(iii) *Ootheca globosa* Kidston

"Probably the microsporangia of Pteridosperms" (Kidston).

(iv) cf. *Coseleya* sp.

"On a piece of shale (V16888) are three groups of sporangia of the same type as Kidston figured. *Coseleya* is otherwise known only from the Westphalian and it seems improbable that the Foulden specimens really belong to that genus".

(v) Fructification of a Pteridosperm

"Some specimens of a much larger fructification, and of some seed-like bodies, will not be described in detail here, since there are numerous better examples in the Kidston Collection, which will doubtless be described in a forthcoming volume of the Survey Memoirs, when Kidston's work is completed".

(vi) *Lepidodendron* sp.

"A single fragment of a stem showing structure (V168720) apparently belongs to the genus *Lepidodendron*. It is interesting as evidence of the occurrence of petrified material at Foulden. Impressions of Lepidodendroid twigs (V16872) occur in a coarse sandstone matrix, and in the shale are isolated megaspores (V16871) like those of *Lepidostrobus* with capitate appendages".

In 1931 Dr Crookall figured sections of *Lyginorachis papilio* Kidston from Norham Bridge (Crookall 1931). The specimen was discovered by Dr Kidston and was first described by Dr Scott (Scott 1923, pp. 57–59). The genus *Lyginorachis* was erected by Dr Kidston (1923, p. 18) for isolated petioles of Pteridosperms having characters similar to those of the petiole of *Lyginopteris* (formerly known as *Rachiopteris aspera* Will.) The petiole of *Lyginorachis papilio* measures 8 × 6 mms. in cross section, and is flattened on what was probably the upper side. It contains a large U-shaped bundle concave upwards and with about ten protoxylem groups on the convex side. The tracheids bear multiseriate bordered pits. The outercortex has the *Dictyoxylon* type of fibrous network. Dr Scott considered that the petiole has more in common with *Lyginopteris* than any other known genus. Dr Crookall also figured a smaller and simpler petiole from the Langton Burn, near Gavinton, but left it incompletely named as *Lyginopteris* sp.

In 1935 Dr Mary G. Calder described two new species of *Lyginorachis* from the West of Scotland (Calder 1935). One of these, *Lyginorachis Waltoni* occurred on the Isle of Arran. It is of interest that further specimens have now been found in Berwickshire at Langton Glen, Hutton Bridge and the Ladykirk Burn.

Dr Calder also described two species of *Lepidodendron* from the Langton Burn, near Gavinton (Calder, 1934 pp. 118–122). One of these, *Lepidodendron brevifolium* Will., has a medullated stele, and was originally described from Burntisland. It has often been referred to the impression species *Lepidodendron Veltheimianum* Sternberg. The other species has a solid protostele, and most specimens show no secondary xylem. This species is very common in Berwickshire, and is unusual in that it lacks clearly defined leaf cushions. Recently it has been re-investigated along with specimens from Arran and re-named *Levicaulis arranensis* (Beck 1958). The smallest axes have been shown to have the leaves attached directly to the stem surface without any evidence of leaf cushions. Furthermore, no ligules have yet been observed at the leaf-bases. Although Beck figured a tangential section through the outer cortex, he had no specimen showing the external features of a petrified stem. Such specimens seen by myself in Berwickshire, show narrow elongated diamond-shaped areas similar to the surface appearance of *Lepidodendron*. It is possible therefore that *Levicaulis arranensis* represents a primitive forerunner of the typical form of *Lepidodendron*, having the leaves borne on non-projecting areas similar in outline to true leaf cushions.

Perhaps the most interesting fossils described by Dr Calder from the Kidston collection of Fossil Plant Slides, are two seeds named

Calymmatotheca kidstonii and *Samaropsis scotica*. Both these I have re-investigated from new material from several localities in Berwickshire as well as from the Langton Burn, near Gavinton, where most of Dr Kidston's specimens originated. It is hoped to publish a full description of these seeds elsewhere. *C. kidstoni* is to be re-named *Genomosperma kidstonii* (Calder) and is of interest in possessing a free nucellus (or megasporangium) surrounded by an integument consisting of eight free lobes which diverge at the apex. A second species *G. latens* is very similar, but has the integumental lobes joined for a short distance at their bases and convergent at their apices, where they simulate a micropyle.

Samaropsis scotica Calder is a platyspermic seed of Pteridospermous affinity, and possesses a wide funnel-like salpinx between two diverging apical horns. It appears to be identical with Kidston's compression seed *Samaropsis bicaudata* (originally named *Cardiocarpus bicaudatus*), which he obtained at Edrom and below Allanton.

Samaropsis scotica is frequently associated with the stem *Stenomyelon tuedianum* and large petioles known as *Kalymma tuediana* Calder, also described by Dr Calder from the Langton Burn, Norham Bridge and Edrom (Calder, 1938, pp 312–329). I have evidence which suggests that *Kalymma tuediana* is the petiole of a large frond borne on *Stenomyelon tuedianum*. This agrees with the Calamopityean affinity of *Stenomyelon*. It would be of still greater interest if *Samaropsis scotica* should prove to be the seed of *Stenomyelon* since at present we have no knowledge of any fructification belonging to the Calamopityeae.

In 1953 W.G. Chaloner described the megaspores from a new species of *Lepidostrobus (L. allantonense)* from the Kidston Collection in the Geological Survey Museum, London, (Chaloner, 1953). The material was collected from the right bank of the Whiteadder, one mile east of Allanton at the locality known as 'Willie's Hole'. Chaloner identified the megaspores with the dispersed spores known as *Triletes crassiaculeatus* Zerndt (sensu Dijkstra 1946). The cone itself is 11 cm. or more in length and 12–16 mm. diameter. The megaspores have a mean diameter of 1.383 mm. and possess an apical prominence and spines typically 200 in length. The spines taper to a fine point; smaller subsidiary spines 35 long are also present. The megaspore wall is typically 200 μ thick. The microspores are unknown. Chaloner suggested that *Lepidostrobus allantonense* may be the cone of *Lepidodendron nathorsti* Kidston.

In 1958 Chaloner described some dispersed spore tetrads from two coals outcropping in Cove harbour and elsewhere. These tetrads named by Chaloner *Didymosporites scotti*, consist of two large fertile spores and two minute abortive spores. They were extracted from the coal by maceration in Schulze's solution (saturated potassium chlorate in concentrated nitric acid) for several hours. After maceration the acid solution was decanted, and the coal washed, then treated with dilute sodium hydroxide solution. Finally the residue was washed, and separated into size grades by sieving. The interest of these spores is that they agree with those occurring in the megasporangia of the primitive fern *Stauropteris burntislandica* described from Pettycur. Formerly these sporangia were given the name *Bensonites fusiformis* and were regarded as glandular bodies. Chaloner's discovery shows that *Stauropteris burntislandica* was among the plants which formed the Cove coals. In addition I have found the plant in a petrified condition in four other Berwickshire localities viz: Langton Glen, Whiteadder below Chirnside Bridge, near Hutton Bridge and on the Blackadder below Allanbank Mill. As yet the stem of *Stauropteris* has never been discovered and some palaeobotanists have doubted whether it possessed one. It is therefore a fossil plant worthy of further investigation.

One other line of research which is only in its infancy so far as Berwickshire is concerned is the investigation of the plants present in peat deposits. As long ago as 1835, David Milne had noted the remains of trees – mainly Birch and Hazel – in the peat mosses of Whiterigg, Whitburn and Dogden. In 1948 G.F. Mitchell described late glacial deposits at Whitrig Bog, which lies 6 miles west of Kelso at 500 feet. Formerly a brick and tile works had extracted clay at the western end of the bog and it was near here at the N.W. margin, that Dr Mitchell and Dr Godwin obtained plant samples from the marls and clay below the peat. Among the plant remains discovered were: *Betula nana* (Dwarf Birch), *Thalictrum alpinum* (Alpine Meadow Rue), *Salix herbacea* (Least Willow) and *Salix reticulata* (Reticulate Willow). These are typical Arctic-Alpine plants now absent from Berwickshire, and they indicate the type of flora which prevailed in late glacial times. Some of the plant remains from Whitrig Bog are figured by Dr Godwin in his book *The History of the British Flora*. Plate XV, and Pl. XXVI. There is little doubt that similar research in other bogs such as Gordon Moss, Penmanshiel Moss and Jordon Law Moss, would produce similar interesting results. What is true for these recent deposits is still also true of the more ancient rocks; new species probably await discovery, and whenever our rivers run in flood, it is possible that new specimens will be uncovered or new strata laid bare for the observant naturalist with an eye for such things.

REFERENCES
(arranged in chronological order)

1843 Milne, David. "Geological Account of Roxburghshire". *Trans. Roy. Soc. Edin.*, Vol. XV, pp. 433–502.

1864 Jeffrey, Alex. *The History and Antiquities of Roxburghshire and adjoining Districts.* Vol. IV., Second Edition, Preface. Edinburgh.

1886 Bennie, J. and Kidston R. "On the Occurrence of Spores in the Carboniferous Formation of Scotland" *Proc. Roy. Phys. Soc.* Vol. IX., pp. 82–117.

1901 Kidston R. "Summary of Progress of the Geological Survey for 1900" (*Mem. Geol. Survey*) pp. 174–175.

1902 Kidston R. "Summary of progress of the Geological Survey for 1901" (*Mem. Geol. Survey*). pp. 179–180.

1903 Kidston, R. "Summary of Progress of the Geological Survey for 1902". (*Mem. Geol. Survey*) pp. 135–137.

1910 Solms – Laubach, Graf Zu. "Uber die in den Kalksteinen des Culm von Glatzisch-Falkenberg in Schlesien erhaltnen structurbietended Pflanzenreste". *Zeitsch. Bot., Jahrg.* II., Heft. VIII, p. 537, Pl III, Figs. 7, 11, 13.

1911 Bertrand, P. "Structure des stipes d'*Asterochloena laxa*. Stenzel". *Memoires de la Société Géologique du Nord*, Tome VII, I, p. 55.

1911(b) Bertrand, P. *L'étude anatomique des Fougères anciennes et les problémes qu'elle soulève*, p. 258.

1912 Kidston, R. and Gwynne-Vaughan, D.T. "On the Carboniferous Flora of Berwickshire", Part 1, "*Stenomyelon tuedianum* Kidston", *Trans. Roy. Soc. Edin.*, Vol. XLVIII, pp. 263–271.

1912 Gordon, W.T. "On *Rhetinangium arberi* a new genus of Cycadofilices from the Calciferous Sandstone Series", *Trans. Roy. Soc. Edin.* Vol. XLVIII, p. 814.

1917 Seward, A. C. *Fossil Plants*, Vol III, p. 205.

1921 Craw, J. H. "Report of Meetings", 1921, *H.B.N.C.*, Vol. XXIV, p. 287.

1923 Scott, D. H. *Studies in Fossil Botany*, Pt. II, 3rd ed.

1923–25 Kidston, R. "Fossil Plants of the Carboniferous Rocks of Great Britain", (*Mem. Geol. Surv.*)

1924 Horne, John. "Obituary notice of Mr A. Macconachie", *Trans. Edin. Geol. Soc.*, Vol. XI, pp. 395–397.

1924 Scott, D. H. *Extinct Plants and Problems of Evolution*; Macmillan and Co. London; p. 162

1927 Hirmer, M. *Handbuch der Palaobotanik*, Vol. 1, München und Berlin; p. 519.

1927 White, E. I. "The Fish Fauna of the Cementstones of Foulden Berwickshire". *Trans. Roy. Soc. Edin.* Vol. LV p. 255–287.

1927 Sanderson, W. (Editor). "The Late Thomas Middlemiss Ovens". *Border Magazine* Vol. XXXII, No. 382, pp. 145–147.

1929   Krausel, R. and Weyland, H. "Beitrage zur Kenntniss der Devon flora", *Abh. Secken Naturforsch, Ges.*, Vol. XLI, lief. 7.

1930   Holden, H. S. "On the structure and affinities of *Ankyropteris corrugata*". *Phil. Trans, Roy. Soc.*, B. Vol. CCXVIII, pp. 79–114.

1931   Crookall, R. "The genus *Lyginorachis* Kidston". *Proc. Roy. Soc. Edin.* Vol. LI pp. 27–34.

1934   Calder, M. G. "Notes on the Kidston Collection of Fossil Plant Slides: No. VI, on the structure of two Lepidodendroid stems from the Carboniferous Flora of Berwickshire". *Trans. Roy. Soc. Edin.*, Vol. LV III. pp. 118–124.

1935   Calder, M. G. "Further Observations on the Genus *Lyginorachis* Kidston", *Trans. Roy. Soc. Edin.*, Vol. LVIII, pp. 549–559.

1936   Read, C. B. "The Flora of the New Albany Shale, Pt. 2, The Calamopityeae and their relationships", *U.S. Dept. Int. Geol. Surv. Prof. Paper.*, 186, E. Shorter contributions to general geology pp. 81–104.

1938   Calder, M. G. "On some undescribed species from the Lower Carboniferous Flora of Berwickshire; together with a note on the Genus *Stenomyelon* Kidston". *Trans. Roy. Soc. Edin.* Vol. LIX pp. 309–331.

1938   Crookall, R. "The Kidston Collection of Fossil Plants. With an account of the Life and Work of Robert Kidston". (*Mem. Geol. Survey*).

1941   Walton, J. "On *Cardiopteridium* a genus of fossil plants of Lower Carboniferous age, with special Reference to Scottish Specimens", *Proc. Roy. Soc. Edin.* B., Vol. LXI, pp. 59–66.

1948   Mitchell, G. F. "Late Glacial Deposits in Berwickshire". *New Phytol.*, 47, p. 262.

1953   Chaloner, W. G. "On the Megaspores of Four Species of *Lepidostrobus*", *Ann. of Bot.* N.S. Vol. XVII, pp. 273–291.

1956   Godwin, H. *The History of the British Flora*; Cambridge pp. 19, 307, 315.

1958   Beck, C. B. "*Levicaulis arranensis* gen. et sp. nov., a Lycopsid axis from the Lower Carboniferous of Scotland". *Trans. Roy. Soc. Edin.*, Vol. LXIII, pp. 445–456.

1958   Chaloner, W. G. "Isolated Megaspore Tetrads of *Stauropteris burntislandica*". *Ann. of Bot.*, N.S., Vol. 22, pp. 197–204.

Appendix 1
Two Sermons

A Sermon, preached at Allanton, December 1946, Chirnside 1946, Cumledge Mill 16.2.1947. Hawick Home Mission August 1947, Douglas Water 1.8.1948, Boston Church, Duns, April 1949, Bonkyl Parish Church, May 1949, Polwarth February 1950, Kelso Baptist Church, 7.8.1960, Leitholm and Eccles 31.12.1961, Scot's Gap 21.7.1968.

> SUBJECT: Divine Antidotes to trouble
> TEXT: John, 14:1. Let not your heart be troubled: believe in God, believe also in me.
> READING: Job, Chapter 3 – end and John Chapter 14.

Introduction: One of old said, "Man is born unto trouble as the sparks fly upward". So we feel we are on safe ground when we preach a sermon on trouble for who among us does not know something about trouble? You will notice that the words of our text are repeated in verse 27. This double use suggests that Christ having uttered the words in the first place goes on to show why his disciples should not be distressed or agitated. The word troubled is used to describe the agitation of water. In between these two exhortations of verses 1 and 27 the Lord gives four reasons his disciples should not be troubled. He shows that there are Divine antidotes provided for every kind of trouble. The first is Believe in God, the second is Faith in Christ. The third is found in verses 2 and 3 in the form of an assurance and promise. "In my Father's house are many mansions and if it were not so, I would have told you, I go to prepare a place for you. And if I go and prepare a place for you I will come again, and receive you unto myself; that where I am, there ye may be also." The fourth is also a word of promise and we find it in verse 16: "I will pray the Father, and he will give you another Comforter, that He may abide with you for ever."

Now we can classify troubles under four heads and if we do so we shall see that these four messages meet them all.

First we have the problems of Creation and providence and the answer here is, "Believe in God". Second we have the problems of sin and Salvation and the answer to these is, "Believe also in Me". Third we have the problems of death and the hereafter, and the answer of Christ is, "In my Father's house there are many mansions". Fourth we have the problem of daily living and conflict with the world, the flesh and the Devil. The answer of Christ to this is, "I will pray the Father and He shall give you another Comforter."

Now we are not all troubled along the same lines but we are all troubled at one time or another with one thing or another. Even royalty are troubled for the saying goes, "Uneasy lies the head that wears the crown".

Some are greatly troubled by the problems of Creation and providence especially since the work of Charles Darwin showing creation by a slow process of evolution by natural selection. The only answer here is, "Believe in God".

Perhaps, however, to most of us the dealings of providence may be a greater trouble to us. Those of you who have read *The Bible in Spain* by George Borrow will remember the tragic story with which he opens his book:

> On the morning of the 10th November 1835 I found myself off the coast of Galicia, whose lofty mountains, gilded by the rising sun, presented a magnificent appearance. I was bound for Lisbon. We passed Cape Finisterre, and standing farther out to sea, speedily lost sight of land. On the morning of the 11th the sea was very rough and a remarkable circumstance occurred. I was on the forecastle, discoursing with two of the sailors. One of them who had just left his hammock, said – "I have had a strange dream, which I do not much like", for continued he, pointing up to the mast, "I dreamt that I fell into the sea from the cross trees".
>
> He was heard to say this by several of the crew beside myself. A moment after, the captain of the vessel perceiving that the squall was increasing, ordered the top sails to be taken in, whereupon this man, with several others instantly ran aloft. The yard was in the act of being hauled down, when a sudden gust of wind whirled it round with violence, and a man was struck down from the cross trees into the sea, which was working like yeast below. In a few moments he emerged. I saw his head on the crest of a billow, and instantly recognised in the unfortunate man the sailor, who a few moments before, had related his dream. I shall never forget the look of agony he cast whilst the steamer hurried past him. The alarm was given, and everything was in

confusion. It was two minutes at least before the vessel was stopped, by which time the man was a considerable way astern . . . A boat was lowered . . . and had arrived within ten yards of the man who still struggled for his life when I lost sight of him; and the men on their return said that they saw him below the water, his arms stretched out and his body apparently stiff but that they found it impossible to save him.

The poor fellow who perished in this singular manner was a fine young man of 27, the only son of his widowed mother. He was the best sailor on board and was beloved by all who were acquainted with him. Yes there is a problem of providence. Why good men suffer and bad men seem to prosper. Jesus suffered and died on the cross and Barabbas was released. Certain things come to one and not to another. Ease to one and constant struggle to another. Health to one and affliction to another. Of course these inequalities often have a very obvious explanation yet at other times, before this strange mixture in life we stand perplexed and baffled and seriously troubled in mind.

Some of you will have been inside a weaving shed and stood before a loom. You will have observed all the separate threads of different hue and tint. How meaningless they seem viewed separately. Some light, some dark, some grey, some gay. We need to see the finished article before we make our judgement, then we see the design. So the word teaches us that, "All things work together for good to them that love God". Believe in God. There's a star to guide the humble. Trust in God and do the right. Though the threads of life may seem dark and meaningless we shall yet see their place in our perfected life.

Remember, you cannot swim on dry land, faith swims in a sea of trouble, and sorrow and perplexity. You cannot have faith without doubt.

>Encamped along the hills of light
>Ye Christian soldiers rise,
>And press the battle e'er the night
>Shall veil the glowing skies.
>Against the foe in vales below.
>Let all our strength be hurled.
>Faith is the victory we know,
>That overcomes the world.
>Faith is the victory,
>Faith is the victory,
>Oh glorious victory,
>That overcomes the world.

> Believe in God.
> Judge not the Lord by feeble sense,
> But trust him for His grace,
> Behind a frowning Providence,
> He hides a smiling face.
>
> Blind unbelief is sure to err,
> And scan His work in vain,
> God is his own interpreter,
> And He will make it plain

Suffering and trouble are part of the price of our freedom. By this gift God puts us on our honour that come what may we will trust him. Therefore to all the problems of creation and Providence this is the answer and antidote.

"Believe in God".

A second class of problems is that of sin and salvation. All men have trouble here. All minds are disturbed and agitated by sin. Every heathen altar testifies to this. We have all tried in various ways to solve these problems. Some do it by giving as little attention to them as possible. They bury their head in the sand and hope for the best. If we are to be saved there has to be repentance and faith, a forsaking of sin and trust in the Lord Jesus Christ.

Some people are like Naaman who came to the Prophet with preconceived ideas of how to be healed of his leprosy. "I thought," he said, "he would surely come out to me and strike his hand over the place." But the prophet didn't do that, he simply said, "Go, and wash, and be clean." The simplicity of God's salvation was turned into a difficulty. The answer to the problem of sin and salvation is in the command of the Gospel. "Believe in the Lord Jesus Christ and thou shalt be saved."

A third problem which sooner or later comes to everyone is the problem of death and the hereafter.

Most of us have been in the death chamber or stood at the open grave.

Here is Christ's answer to those questions that come to us.

"In my Father's House are many mansions . . . I go to prepare a place for you." This gives us the assurance of eternal life. Therefore, let not your heart be troubled. In my Father's house there are many abodes. Jesus himself lay in the grave and said beforehand, "When I am raised I will go before you," and "Because I live, ye shall live also." Meanwhile we walk by faith and not by sight.

The fourth and last class of problem is that of experience as children of God. As believers we soon discover that we are not exempt from the fiery trial and like Daniel we discover that there are lions allowed to cross our path. Though like him we pray three times a day temptations are as keen if not keener than before we were converted. Christ's answer to this confict is, "I will pray the Father and He shall give you another Comforter that He may abide with you for ever." The disciples were anticipating a terrible and lonely experience. For such there is the blessed gift of the Divine Presence. He is promised to each believer. As at Pentecost there was an individual tongue of fire which abode on the head of each disciple, man or woman, so to every believer there is promised the personal gift of the Spirit of God. Therefore let not your heart be troubled.

> The Divine Presence is with you.
> The Divine Sanctifier is in you.
> The Divine Power is over you.
> Let not your heart be troubled.
> Believe in God,
> Believe in Christ,
> Believe in the Father's Home
> Believe in the Holy Spirit
> And the peace of God which passeth all understanding shall keep your hearts and minds through Christ Jesus.
>
> Amen.

A Sermon: preached at the Faith Mission 600th conference, Edinburgh, Carrubber's Close, 1 December 1945, Kelso Baptist Church, 6 January 1946, Allanton, 8 January 1947, Chirnside Parish Church and Edrom, 30 June 1963.

SUBJECT. The life of Faith.
READINGS, Genesis Chapters 12 and 13.
TEXT: 2 Corinthians 5:7. For we walk by faith and not by sight.

Introduction:
In this text Paul asserts that the Christian life is a life of faith. The emphasis on faith is not peculiar to Paul. It is one of the unique features of the Bible and especially of the New Testament.

Again and again the importance of faith is stressed and we read words like these:

> Where is your faith?
> Have faith in God.
> Thy faith hath saved thee.
> According to your faith be it unto you.
> Only believe.
> If thou wouldest believe thou shouldest see the glory of God.
> This is the work of God that ye should believe on Him whom He hath sent.

The fact that Paul spoke these words suggests that we need to remind ourselves of this truth.

We would all prefer to walk by sight rather than by faith.

The fact that the Christian life is a life of faith is a sure sign of the goodness and wisdom of God:

First.

It is clear that if we are to possess true freedom we cannot possibly know the future. That would give us no choice in the matter. Instead God has given us freedom and trusted us to trust Him and obey.

Second.

The Divine goodness is also shown in that faith places salvation within the reach of all who hear.

Third.

Salvation by faith ensures that all the glory shall be given to God which is as it should be.

Now after Christ there is no greater example of the life of faith in the Bible than Abraham. "Abraham believed God". His life is an illustration of what it is to walk by faith and not by sight. It shows us how such a life begins, continues, and what strange experiences it may meet.

I The life of faith begins in a Divine call, a self-discovery and an act of obedience.

Where was Abraham when God called him? He was in Ur of the Chaldees, a land which Jeremiah tells us, was mad upon idols. A divine revelation was given in the form of a command. "Get thee out and I will bless thee."

The Christian life begins not in self but in God, in a seeking Saviour and a Divine call. The call of the Gospel is a command and a promise. Believe on the Lord Jesus Christ and thou shalt be saved.

This is how the Christian life begins, by faith in Christ. This is the new birth which begins the new life.

A self-discovery.

When God found Abraham he found him in a wrong place and a wrong condition living among idols – an idolater. In like manner when the Gospel of grace found us we were in a wrong place and condition. A fourfold description of our state by nature is given in Paul's epistle to the Romans Chapter 5.

Here we are described as being,

> Without strength unable to do what we would,
> Ungodly i.e. the reverse of godly
> sinners i.e. transgressors of God's law
> enemies i.e. opposed to God.

The light of this revelation discovers to us our unspeakable need of a Saviour.

An act of obedience.

We read in the epistle to the Hebrews, chapter 11, verse 8, "By faith, Abraham when he was called . . . obeyed, and he went out not knowing whither he went". The promised blessing was conditional upon this obedience.

The Christian life likewise begins in the obedience of faith.

Faith without obedience is no faith at all. If you and I REALLY TRUST in Christ we shall rise up and follow Him. We shall walk by faith and not by sight.

II This life of faith may be a life of partial and incomplete obedience.

Where was it that God called Abram to when he called him out of Ur? It was to a land that God would show him – Canaan.

And where was Haran?

It was a town on the Border. Betwixt and between the two countries. Neither one thing nor the other. On the one hand he could see the country he had left, on the other hand he could see the country he should have entered. What was it that kept Abram in Haran? It was a natural tie. We read that it was not until his father Terah had died that "Abraham departed as the Lord had spoken unto him".

What does Canaan represent in the Christian life?

It represents the full inheritance of what God promises in this life. It is the fulness of the Holy Spirit, the seal and the earnest of our inheritance in heaven.

III A life of Growth and progress.

In Genesis chapter 11, verse 31 we read "Abram went forth". Thereafter we read phrases as

> Abram departed.
> Abram passed through.
> Abram journeyed.
> Abram went down.
> Abram went up.
> Abram dwelt.
> Abram removed.

All suggest that the life of faith is a life of growth and progress.

It was true that Canaan was Abram's as soon as he embraced God's promise nevertheless he did not know one tenth of what it contained for him. It is the same in the Christian life. All the blessings of Salvation are ours the moment we believe but the knowledge of them is progressive. A little girl may receive the gift of a purse. All is hers the moment she receives it, but she does not know its worth and wealth fully. And so it is with Christ and His redemption.

"Grow in grace and in the knowledge of the Lord Jesus Christ". May we be saved from settling down. That is the way to stagnate. It is moving water which keeps its sparkle and freshness. The life of faith is one of progress and growth right to the end. "Abram went forth".

IV It may be a life of strange and trying experiences. If there were no problems where would be the province of faith? Faith is like swimming. We must take our feet off the bottom. It is when a man is pushed into water that we see whether he can swim. So it is with faith. Its province is difficulty, uncertainty and jeopardy. So it was with Abraham.

> Genesis 12:6 "The Canaanite was then in the land".
> Genesis 12:10, "There was a famine in the land and Abram went down into Egypt."
> Genesis 13:4. Abram retraces his steps, "Unto the place of the altar which he had made there at the first."
> Genesis 13:7. "And there was a strife between the herdmen of Abram's cattle and the herdmen of Lot's cattle."

And Abram said unto Lot, "Let there be no strife. I pray thee between thy herdmen and my herdmen for we be brethren."

"Is not the whole land before thee? Separate thyself I pray thee

Appendix 1

My wife Gladys with Dr H. B. M. Auld, our former much esteemed G.P. at Duns, Berwickshire. Photograph taken on our Golden Wedding Anniversary, 20 June 1992, by Siobhan McDermott.

from me. If thou will take the left hand then I will go to the right hand or if thou depart to the right hand, then I will go to the left."

And Lot lifted up his eyes and beheld all the plain of Jordan that it was well watered everywhere.

Then Lot chose him all the plain of Jordan. He pitched his tent toward Sodom. And the Lord said unto Abram after that Lot separated from him. "Lift up now thine eyes and look from the place where thou art northward, and southward and eastward and westward. For all the land which thou seest, to thee will I give it."

A man of faith can always afford to wait.

Faith may not be quick but it is sure. Faith was justified and Lot was the loser in the long run. He had pitched his tent toward Sodom. Before long he was in Sodom and then Sodom was in him and he only escaped destruction by the skin of his teeth and the intervention of the grace of God through his servant Abram. Yes Abraham believed God and it was reckoned unto him for righteousness. Now this was not written for his sake alone but for our sake also, unto whom it shall be reckoned who believe on him that raised Jesus our Lord from the dead, who was delivered up for our trespasses and raised for our justification.

Appendix 2

SOME LETTERS THAT HAVE AN INTEREST FOR ME.
COPY

Egypt, 28 Nov. 44.

Dear Mr Long,

The re-appearance of a *Lagenostoma* once more in the Annals, under your hand, interests me. This species (*L. ovoides*) is by far the commoner in the coal-balls and likewise often well provided with pollen. In view of its frequency, from an early stage in the *Lagenostoma* work, I had Lomax sending me all the specimens that came his way. The particular interest at the time was in the cupule, but in all the *ovoides* I looked at (perhaps 100) the seeds had fallen from their husks (anyway I never got continuity). Much later I handed the whole lot over to Miss Prankerd, who wrote about them in the Linnaean Journal in 1912. All fossil seed preparations obtained by me are in the U.C.L. collection, and if they escaped blitzing should be available when U.C.L. has sorted itself out and returned to its proper home in Gower Street.

The odd section of *L. lomaxii* handed to McLean turned up long after the Oliver and Scott paper of 1904, and as a Lycopod megaspore with prothallus arrived at the same time, McLean also had that. He was a splendid draftsman and the New Phytologist (1912) did full justice to his efforts.

I myself began looking at fossil "seeds" of Carboniferous age as early as 1897. They were always detached, and at the same time many "ferns" were deficient in known reproduction. There was the evident possibility that these two circumstances were related. It was a detective job.

From an examination of conifers in living collections I found in many cases (notably in *Torreya californica*) that ovular stages were often cast in immaturity, so that possibly the evidence I was in search of might be discovered in connection with precocious abscissions.

Williamson, had he devoted himself to this objective, should by right have discovered "Pteridosperms" at least 15 years before Scott and I got hot on the scent. But his hands were so full of more complete and attractive material that he was rather careless of *disjecta membra*, and on more than one occasion would describe transverse and longitudinal sections of the same object as distinct and unrelated things.

My numerous visits to France – Paris, Autun, Rive de Gier etc. drew a blank in this respect, as the French silicified material, though beautiful, is too fragmentary. However, I worked out *Stephanospermum*, and it was one of the specimens of this seed from the S. of France that provided the rare occasions of archegonia in a plausible position (these also are in the U.C.L. collection).

Walton's "peel" technique was not available in those days, and as a rule a seed was only revealed when the lapidary's saw went through it. One was lucky to get 3 or 4 transverse sections of those little seeds. Moreover Lomax, should he get 3 sections through one seed had the annoying habit of sending each to a separate client. He thus gratified 3 clients at once, and could charge a higher price. In this I was once victimized - it was a long series showing branching in *Lyginopteris*. Noting the gaps to be unduly wide I discovered in time that the alternate sections had gone to the Bertrand collection and the Bertrands behaved well and released their specimens. After that Lomax was subjected to a more severe discipline, as a group of British palaeobotanists had found capital to extend his workshop and equipment, and had the whip hand.

Lomax who was turning out sections by the thousand threw away unsaleable "rubbish" ie. such sections as would not fetch a price. These rejects I found chucked anyhow into crates and bought them at 2/6 a hundred. He would send a thousand at a time. I looked them all over with a hand lens first and then with a microscope, and obtained quite a number of overlooked good things that way. For instance, our R. 13 (= fig. 34 on Pl. 10 of Oliver and Scott's *Phil. Trans.* paper of 1904) turned up that way, and a valuable element it proved to be when we learnt how to interpret it. One of the 2 preparations described by McLean turned up in the same way – I forget which of them.

The *Conostomas* were also interesting. For ten years Lomax was collecting them before Salisbury and I had enough to justify publication. It was in connection with that paper that we made a "megatome" – which consisted of an adjustable frame which could be set at any desired angle around a wax model – say of a seed held vertically – sliding on which frame we employed a wire with a handle at each end and cut the model like a grocer cuts cheese. Such practice in the interpretation of planes of cutting of oblique sections is invaluable in fossil work. For instance in *Annals of Botany* (1906 or 7) Scott and Maslen describe a new species of *Trigonocarpus* as being a "coffin shaped" seed ... Actually this seed is a perfect hexagon in transverse section, as Salisbury and I (or Salisbury) discovered later. Now Scott was a man who rarely or never stumbled, and in this instance doubtless the misinterpretation must have been due to Maslen. I hope I don't do him an injustice.

Your stuff I see was from Dulesgate – a locality we always respected. If post-war there should be reconstruction and reorientation in coal mining technique, and the opening of new seams. It should be a great opportunity for the palaeobotanist. Virgin localities are apt to yield new mixtures, new forms and types of preservation differing in certain respects. Manchester is conveniently situated in relation to such localities, and it might be worth while for somebody in the Department to acquire the habit of regularly visiting such places. I wonder is Lomax fils still connected with coal research? Lang of course who attained such great distinction in fossil botany, always kept close contact with localities – often in remote and inaccessible places - but he needs no bouquets from a "foreigner" like me. With apologies for all this screed,

from yours sincerely
F.W. Oliver

Dar el Nabati,
Burg el Arab,
Mariut.

COPY

Egypt. 24 Feb. 1945.

Dear Mr Long,

I have to thank you for separate of your *ovoides* paper. I note that you designate the preparations by numbers 1, 2, 3, etc. and the specimens A, B, C, etc.

Presuming that "peels" are permanent, it is vital that the critical preps should be given final numbers in some registered collection (e.g. Manchester) and there stored. As you would see in Oliver and Scott, 1904, we had some trouble in assembling dispersed sections of *L. lomaxi* owing to the vicious practice of Lomax père, when he had but 3 sections of one seed, of sending each to a distinct collector. The only fossil paper I have here is Oliver "on *Physostoma*" *Ann. of Bot.* 1909, where at pp. 80 and 113, you will find the system illustrated. In fossil work such precision is the more important in view of the probable long intervals between successive publications relating to the same species. Please pardon the admonition.

Following the establishment of Lomax as a company, Lomax was still liable to disperse sections – especially of new finds. Scott and I raised the matter at a Director's meeting. Sutcliffe, a cotton spinner at Shore (Rochdale) and the owner of the Shore mine, was also a Director, and he supported Lomax vigorously in this practice. His argument was "This is a commercial company and it is up to Lomax to get the best prices he can". Morally, of course it wasn't a commercial company. However, we downed Sutcliffe and the objectionable practice ceased.

In those days the utmost good fellowship prevailed among workers in this field. I recall how Vienna, Paris, Berlin loaned us unique specimens, whilst we, reciprocally, induced the Trustees of the British Museum to suspend their previously inflexible rule that once a specimen had found a place in the B.M. collections, it was irremovable.

Where you are now living must be classic ground in palaeobotany, Black and Whiteadder; Lennel, etc. favourite hunting grounds of Kidston's. In 1881, when I hardly knew there was such a realm as fossil botany, I remember spending a night at Norham, and next day walking to Coldstream, Kelso and Melrose.

Kidston once told me he had picked up quite useful specimens from broken up road material and from walls in that district.

Though I have visited the Northumbrian and Scottish coastline as far as the Firth of Forth, I know best W. Central Northumberland – Woodburn just S. of Otterburn, where my forebears farmed up to 1760.

8 or 9 years ago I re-explored the district, and judge it to have altered little in the last 200 years. I fancy that enclosure acts, whereby we were denied grazing rights on the heights, must have induced a certain Andrew Oliver (a key man in the ancestral line) to sell the farm and settle much nearer to Newcastle-on-Tyne at Benwell Hills - a property that remained in

the family till 1850. It is all built over now through the spread of Newcastle. However, I have somewhere a drawing of the district circa, 1841, with the aid of which I might be able to identify the site, should opportunity serve.

You are quite right about the rarity of the seed *L. lomaxii*. The section that McLean described turned up late, and as the preservation was excellent I handed it to him as he was a good draftsman as well as interested in such objects. Had the section hit another plane we might have had a glimpse of archegonia.

Yours sincerely
F.W. Oliver

COPY

2 Heaton Rd., Withington,
Manchester, 20.

Jan. 8, 1945.

Dear Mr Long,

I was very interested in your letter. I am increasingly out of touch with news or might have heard of your migration North etc. I hope you will find the school and work congenial and the move a success.

I don't know your world personally. The nearest I have been is North Berwick. But you seem about on the border line of the Cementstone group of the Lower Carboniferous and the Upper Old Red Sandstone and there should be interest for you in these and other rocks. All I have seen from the Upper O.R.S. are some ribbed casts from the Whiteadder in the Kidston Collection – perhaps you may find something more determinable. The only definite plants from this formation in S. Scotland are still the specimens of *Archaeopteris hibernica* in the Hugh Miller Collection, Edinburgh.

In the Lower Carboniferous beds of Tantallon S. of North Berwick and at Gullane to the N. there are petrified remains in volcanic ash. This is Gordon's collecting ground. I am sure when you get established you will find it a good world for geology and other interests.

I have read Prof. Oliver's letter with great interest. It is very stimulating to have such first hand reminiscences of a pioneer in response to your paper on the *Lagenostoma* prothallus. I am sure you appreciated getting such a letter. It is returned with this.

My wife and I are both well but rather too much occupied. I am doing very little constructive work- making bricks without new material is not my job – but (I) go in a number of times a week to the Museum.

I shall look forward to seeing you and hearing your reaction to Berwickshire if you are in Manchester.

With kind regards and best wishes for your new post

Yours sincerely
W.H. Lang

COPY

2 Heaton Rd.,
Withington, Manchester 20.

20th May, 1945.

Dear Mr Long,

Your letter and little parcel must have come to the Museum a month ago on the day we had gone on a holiday to the Lakes. Being a "parcel" it was not forwarded and I only got it when I went into my room on Friday.

You have certainly found a fossiliferous bed which will give you some fun. Thanks to your description and sketch map I have been able to find the spot on Geikie's old Geological Map of Scotland which is my general source of reference for regions of which I have not needed to get the Survey maps.

It is quite clear that the stretch of coast from Pease Bay just south of Cove Harbour to a short distance S. of Dunbar is Lower Carboniferous and your bed is probably in the Cementstone Group.

If you have not meanwhile found out this or more, perhaps this will be of use.

Anyhow it will be well worth your working at the bed whether the Geological Survey know of it or not. You could ask them if you are in Edinburgh any time.

I heard from Dr Stockmans (Musée d'Histoire Naturelle, 31 Rue Vautier, Brussels). He was asking after you. If the post will take them, (or when it will) send him copies of your papers. I also heard from Prof. Suzanne Le Clerq, from Liège.

These are the only Continentals I've heard from as yet. I have no news to justify another sheet.

Yours sincerely
W. H. Lang.

Tail Piece
Nature Jottings
Summer Nights

From the Todmorden Public Library Journal. Written about 1933 while at the Todmorden Secondary School.

Slowly but surely slips daylight into darkness, and except for a paling western glow the heavens hang sombre and indistinct unlit by moon or star. A bat flits by on velvet wings and above, just within my range, shriek a company of swifts. Few yet various sounds come to my ears intermittent with the not unpleasant sound of a distant mowing machine. Now and again one hears the quavering call of an owl in the wooded slopes of the valley far below, but the silence of the evening is as enjoyable as the different cries of the wild creatures.

Stoodley Pike, a famous landmark near my home town, Todmorden, first built at the close of the Napoleonic War in 1815 and rebuilt after the Crimean War. Photograph by courtesy of Todmorden Photographic Society.

Tail Piece

Ghost Moths swing pendently over the luscious meadow grass and almost at my feet a map-winged swift noisily beats its wings untiringly against the grasses as it attempts to rise and seek its mate; its efforts are in vain for it is in the death clasp of three murderous black ants. This grim little tragedy to my mind symbolises Satan dragging down mankind, thwarting his ambitions and aspirations, keeping him from the higher life that may be his.

A field mouse which passed by a minute ago, silently as a shadow, is now quite audibly nibbling away at the roots of some such plant as a rush. Something else also engages my attention – it is the hordes of midges round my neck and forehead. Their numbers are "continuous as the stars that shine" so I decide to rise from where I have been sitting and descend from the upland, through the noisy bracken to a pasture on the fringe of a wood. There I am suddenly brought to a halt by a rustling and still louder sniffing. After a few seconds I detect him, an old hedgehog, crouching to the turf, and rolling him over with my stick he becomes a mass of spines from which he will emerge when I am gone to continue his nocturnal search for slugs and beetles.

Brimstone moths flit around the hawthorns, and my approach though silent disturbs a cuckoo, which hawk-like flies to a dead sycamore two hundred yards or so away. On calling thrice he awakens a nearby robin, which after uttering a sweet and abridged version of its song, leaves me again with the solemn stillness of the night. Reluctantly I make my way home with feelings of a mixed tranquillity.

Reader know you the joys of a night ramble, to be appreciated to the fullest extent alone? Have you watched the moon set between the trees while the continual prattling of a brook works a narcotic effect on one? Have you seen the throat feathers of a tawny owl quiver as it utters its trembling call, thrilling the woodland so that the rustling leaves maintain a forced silence? If you have, you have got something that cannot be purchased by pieces of silver. You have a chaste and beautiful memory.

Index

Abbey St. Bathans 75, 77, 85, 96, 101, 104, 107, 159
Aberlady Bay 81
Adiantes antiquus 214
Aeshna 136, 138, 139
Aiken, Rev. J. J. M. L. 175
Aiky Wood 107, 128, 144, 156
Alcicornopteris 190, 213, 228, 229, 234
Allan, P. B. M. 101
Allanbank 95, 193, 205, 209, 211, 212, 232
Allanton 22, 106, 211, 228–30, 234, 235, 241, 245, 249
Allen, A. R. 123
Allendale 161
Alston 117, 118
Ambler, Isabel 3, 7
Amyelon 56, 67
Anabathra pulcherrima 22, 205, 206, 210
Anachoropteris 36, 64
Anachoropteris pulchra 67
Anderson, Adam 103, 123
Anderson, John 103, 123, 141
Andrews, Prof. Henry N. 19, 45, 197, 206
Aneimites acadica 190, 229, 230, 234, 239
Aneurophyton germanicum 232
Angiosperms 16, 18, 167, 169, 172–4, 190–2, 197–201
Angle Shades 143, 156
Ankyropteris corrugata 49, 53, 67, 232, 244
Ankyropteris westphaliensis 56, 67
Annulet 153
Anomolous 137–9, 154
Antler Moth 114, 178
Aphlebia 229

Apperley Dene 161
Araucaria 195, 211, 215
Arber, A. 69
Archaeopteris 170, 171, 183
Archaeopteris hibernica 202, 258
Archaeosperma arnoldii 170, 171, 183
Armstrong College 42
Arnold, C. A. 171, 183
Ash, J. S. 161
Ashworth, Thomas 41, 45, 54
Aspen 179
Asplenium bulbiferum 20
Asterocalamites scrobiculatus 230, 231
Asterophyllites 34
Auld, Dr H. B. M. 253
Autumn Green Carpet 73
Autumnal Moth 120, 142
Autumnal Rustic 139
Avery Hill 5
Avicula 210

Backhouse, William 114
Bacup 6, 11, 34, 39
Balfour, Prof. I. B. 204, 220–3
Bamburgh 112, 113, 181
Bank Hall Colliery 37, 62
Bardon Mill 161
Barker, Jess 8
Barker, R. W. 180
Barnard, P. D. W. 16, 174, 183, 188
Barn Owl 80, 146
Barramill 170
Barred Carpet 122
Barred Chestnut 139, 153
Barred Umber 113
Bar-tailed Godwit 77, 81
Bassendean 107
Bateman, Bertha 7
Beadnell 181

Bearberry 182, 183
Beautiful Brocade 120, 123
Beautiful Carpet 134
Beck, C. B. 170, 171, 183, 240, 244
Bedlington 113, 122, 179
Bedstraw Carpet 73
Bedstraw Hawk Moth 103, 112, 115, 179, 180
Beech Fern 87, 96
Beech Green Carpet 113, 126
Belford 112, 119
Bell Wood 126, 131, 135, 137, 138, 150
Bell's Burn 228, 230, 234
Belling 161
Bennie, James 226, 227, 242, 243
Benson, Robert H. 60, 121, 179, 189
Bensonites fusiformis 183, 242
Benwell 179, 181, 257
Bertrand, P. 183, 232, 243, 255
Berwick 118, 158, 181
Berwick High School 12
Berwick-upon-Tweed 19, 97, 150
Berwickshire High School 12, 14, 16, 97, 184
Berwickshire Naturalists' Club 75, 78, 99, 111, 158, 167, 181, 203, 207, 208
Betula nana 242
Bible Training Institute 5
Bigge, J. F. 160
Billie Mains 181
Binney, E. W. 28, 30, 35, 69, 218, 219
Birch Mocha 113
Birgham House 179
Birkbeck College 16, 32, 45, 188
Birkett, Dr. N. L. 105, 125, 165
Birtley 120
Bishop, J. 237
Black Grouse 91, 145
Black Guillemot 71
Black Rustic 117, 125, 142, 155, 156
Blackadder 22, 83, 142, 206, 210, 211, 226, 228, 242, 257
Blackadder Woods 106
Blackhall Rocks 163

Blackneck 156
Blackpool Moss 164
Black-tailed Godwit 81
Blake Dean 135
Blanerne Bridge 228
Blossom Underwing 112
Bolam 179
Bolam, George 70, 84, 85, 88, 101, 104, 106, 117, 118, 123–7, 134, 147, 150, 158, 160–2, 164, 179, 181
Bolam Lake 161
Bold, Thomas John 114, 115
Bonkyl House 13
Bonkyl Wood 71
Bonnyrigg Hall 161
Bordered Sallow 126, 135, 152
Bordered Straw 113
Borrow, George 246
Botany of the Eastern Borders 100
Bothrodendron 37, 46, 51, 62, 66
Botryopteris antiqua 59, 60, 184, 186
Botryopteris cylindrica 41, 49, 50, 53, 58, 62, 67
Botryopteris hirsuta 20, 30, 31, 37, 48, 51–4, 57, 61–3, 67, 69, 184, 200
Boulger, G. S. 204, 224, 225
Boulsworth 44
Bowhill 164
Bowman, J. T. B. & B. 161
Bowman, R. B. 176
Boyd, W. B. 159, 164
Brachiopods 211
Braithwaite, M. E. 162, 165, 180, 181
Brambling 75, 79, 80, 89
Brick, The 141, 156
Bridgeheugh 164
Brimstone Butterfly 121
Brindled Ochre 106, 142, 145, 146
Brindled Pug 146
British Museum (Natural History) 59, 235
Britten, Mr. Harry 99, 224
Briza media 179
Broad-barred White 150
Broad-bordered Yellow Underwing 137
Brockenhurst 121

Index

Brockholes 176
Broken-barred Carpet 132
Brongniart, Adolphe 207
Broomdykes 158
Broomhouse 103, 149, 187, 211
Broomhouse Mains 158
Brough 223
Brown Argus 105
Brown Crescent 116
Brown Rustic 131
Brown Silver-line 73, 127, 131, 148
Brown Tail 103, 115
Brown, T. 176, 178, 182
Brownlie, Wattie 17
Brunton 162
Brussels Lace 127, 135, 151, 152
Bryson, A. 218
Buckham, A. G. 159, 164,
Buckley Wood 157
Buff-tip, The 110, 150, 152
Buglass, Simpson 88, 103, 107, 123, 125, 154
Bullfinch 97, 98
Bulrush Moth 126, 140, 141
Burleigh, D. G. 181
Burnet Companion 121
Burnet Saxifrage 179
Burnett, John 16
Burnfoot 164
Burnished Brass 109
Burnley 4, 6, 18, 27
Burnmouth 90, 95, 97, 102, 103, 132, 153, 154, 156, 180, 187, 212
Burradon 160
Butterbur Moth 107, 125, 138, 139

Cabbage Moth 140
Calamites 29, 37, 49, 50, 51, 54, 55, 62, 63, 65, 68, 208, 226
Calamopityeae 232, 233, 241, 244
Calamopitys 94, 190, 200, 214
Calamopitys beinertiana 214
Calamostachys 66
Calathopteris 190, 200
Calathospermum 174, 192, 197-9
Calathospermum fimbriatum 173, 183, 191
Calder, Dr. Mary G. 232, 240, 241, 244
Callaly 159

Callixylon 170, 183
Calymmatotheca affine 231, 235
Calymmatotheca kidstonii 94, 183, 189, 241
Camberwell Beauty 102, 114
Cambo 160
Campanula latifolia 179
Camptosperma 190
Canary-shouldered Thorn 127, 141, 142
Carabus nitens 178
Carbonarius 210
Cardiocarpus bicaudatus 229
Cardiopteridium 231, 244
Carex demissa 178
Carex echinata 178
Carex lepidocarpa 178
Carex nigra 178
Carline Thistle 90
Carrubber's Close 249
Carruthers, R. G. 2:3, 217
Cash 44
Catcleugh 118, 119, 161
Cattleshiel 176, 178
Caw Lough 161
Census Catalogue of the B. B. S. 100
Centre Barred Sallow 140
Centre Vale Park 157
Chalk Carpet 113
Chaloner, Dr. W. G. 171, 183, 241, 242, 244
Chamomile Shark 103, 113
Chapman, Abel 118, 160
Cheltenham 2
Cherrytrees 164
Chester-le-Street 163
Chestnut, The 73, 145
Chestnut-coloured Carpet 113
Chiffchaff 95
Chillingham 106
Chinese Character 153
Chirdon Burn 162
Chirnside 245, 249
Christy's Carpet 120
Cinnabar 109
Cladoxylon kidstonii 231-3
Clay, Dr. 177
Clifden Nonpareil 100, 102
Cloaked Carpet 113

Cloaked Pug 119
Close House 162
Clouded Brindle 102, 116
Clouded Buff 104
Clouded Drab 73, 128, 130, 144
Clouded Silver 122
Cloudy Sword Grass 116, 119, 145, 156
Clough, C. T. 203, 213, 214, 217
Cloughfoot 30, 31, 39, 43, 61
Coal-ball localities 6
Coal Law Wood 117
Coccosteus 61
Cochlearia danica 90
Cockburn Bridge 170
Cockburn Law 75, 96, 127, 132, 216
Cockburnspath 20, 185, 212, 213, 226, 230
Coldgate Burn 120
Coldingham 104, 107, 148, 151
Coldingham Moor 23, 100, 132, 149, 158
Coldstream 101, 104, 155, 158, 179, 213, 226, 236, 257
College Valley 85, 185
Collingwood, F. J. W. 211, 217
Comma 100, 103, 105, 109
Common Blue 93, 108, 179
Common Bordered Beauty 121
Common Chestnut 112
Common Ear 138, 143
Common Heath 133
Common Lead Belle 152
Common Quaker 73, 128, 130, 148
Common Sallow 138, 139, 141
Common Scoter 82
Common Shark 113
Common White Wave 73
Confused 118, 137, 138, 154
Coniferae 210, 211, 214, 215
Conostoma 255
Convolvulus Hawk Moth 24, 88, 92, 97, 112, 142, 154
Copper Underwing 116, 123, 157
Corbridge 161, 181
Cordaiteae 207, 214, 215
Cordaites 34, 68, 195
Corixa 136
Cormorant 82, 96

Cornsalad 90
Corsbie Moor 107
Coseleya 234, 239
Cottingham 8, 10
Cove Harbour 20, 153, 185, 226, 230, 231–5, 241, 259
Cowberry 182, 183
Coxcomb Prominent 131, 133, 151
Craigleith 193, 194, 207, 216
Craigs, Robert 118, 119, 120, 161, 179
Cranshaws 70, 104, 107, 126, 138
Craw, J. H. 182, 236, 243
Cream Wave 113
Crescent Dart 113
Crescent Striped 116
Cresswell 160, 161
Cridland, Dr A. 15
Crinan Ear 143
Croft 163
Croft, W. N. 45, 60–3
Cromwell 10
Crookall, Dr. 225, 233, 240, 244
Crooked Burn 195, 196, 234, 235, 237, 238
Crookholm Wood 165
Crossbill 80, 82
Crowberry 85, 89, 179
Croxdale 222, 223
Culley, Matthew 216
Cumledge 22, 80, 94, 104, 171, 245
Currant Spinach 153
Cycadites Caledonicus 213, 226
Cycadofilices 214, 232, 243
Cyclopteris 202, 212
Cystosporites devonicus 171

Dadoxylon 194, 214
Dallman, A. A. 61
Dark Bordered Beauty 118, 120
Dark Brocade 131
Dark Chestnut 112
Dark Dagger 116, 123
Dark Green Fritillary 111
Dark Sword Grass 130, 139, 155
Dark Tussock 133, 144, 151
Darlington 163
Darnell, Rev W. 182
Darwin, Charles, 246
Davidson, Mr. W. 180

Death's Head Hawk Moth 88, 92, 97, 109, 154
December Moth 128, 143, 156
Deep Brown Dart 106, 125, 139, 154
Deerplay Colliery 31, 40, 51
Delevoryas 197, 198
Deltasperma 189, 190
Denholm 164, 165
Dew Footman 90, 132, 145
Diamond Back Moth 97
Dick, Robert 61
Dictyoxylon 240
Didymosporites scottii 241
Dingy Footman 112
Dingy Skipper 112
Dinnington 111, 113
Diplopteridium teilianum 195
Diplotmema adiantoides 235
Diplotmema dissectum 234
Diplotmema patentissima 234
Dipper 75, 79, 86, 91
Dipton 122, 160, 162
Dirrington 78, 127, 178, 182
Doals 11
Dogden 242
Dogden Moss 145, 146
Dolichosperma 190
Dotted Border 71, 76, 144
Dotted Carpet 127, 136, 137
Dotted Rustic 102
Dotterel 148
Double Lobed 113, 121
Double Square Spot 135
Doubleday, Edward 101
Douglas, Dr. Frances 177
Douglas Water 245
Drakemire 149, 176
Drinker, The 105
Dronshiel 178
Duerden 32, 45, 54
Dulesgate 15, 29, 30, 31, 33, 38, 41, 43, 46, 53, 55, 57, 61, 62, 64, 66, 256
Dunlin 77
Dunn, T. C. 163, 175
Duns 12, 13, 17, 75, 104, 143, 154, 158, 184, 245
Duns Castle Lake 24, 76, 140

Dusky Lemon Sallow 101, 113, 126, 154
Dwarf Pug 132
Dytiscus marginalis 136

Eagle's Crag 68
Eales, H. T. 117, 162, 179, 181
Eared Sallow 179
Earlston 100, 103, 104, 159
Early Engrailed 86
Early Grey 117
Early Moth 75, 83, 122, 128, 144
Early Purple Orchis 90
Early Thorn 119, 131, 147
Early Tooth-striped 130, 146
Eccles 159, 211, 245
Eccroustosperma 190
Eckford, R. 203, 204, 217, 236
Edinshall Broch 96
Edrom 94, 104, 107, 157, 174, 181, 228, 230, 235, 241, 249
Edwards, W. N. 60, 235, 238, 239
Eildon Hills 104
Elba 75, 87, 142, 144, 187
Elemore 163
Elliot, Adam 164
Elliot, Miss G. A. 157, 165, 177, 179
Elliot, Sir Walter 164
Elyhaugh 181
Embleton, Robert Castles 176, 182, 209, 216
Embryo Sac 169
Emperor Moth 73, 131, 146, 178
Empetrum nigrum 179
Engrailed 119, 120, 132, 146, 148
Eosperma 190
Erica tetralix 178
Eristophyton beinertianum 214, 233
Etapteris scottii 55, 64, 66
Eurypterid 213, 226
Eurystoma 190, 191
Evans, Arthur 177, 178
Evans, G. 179
Evans, W. 159
Eyemouth 74, 102, 123, 132, 151, 155
Eyre, M. 162

Falconer, Alan A. 178

Fallowlees Burn 120
Fanfoot 126
Fans 104, 105, 123, 176
Fast Castle 88, 134
Fauna of Twizel 112
Fauna Suecica 108
Feathered Gothic 125, 153
Feathered Thorn 127, 143
Fenwick, G. 161
Ferguson, John 106, 175
Ferney Lee Road 5, 7
Figure of Eight 106
Finlay, John 114, 116–18, 123
Flame, The 142
Flame Carpet 135, 141, 142, 148, 149
Flame Rustic 125
Flounced Chestnut 103, 141
Flowerscar 56
Ford, E. B. 91, 105
Forster, Frances 206
Foul Burn Bridge 176
Foulden Newton 196, 234, 235
Four-dotted Footman 121
Fox Moth 106, 145, 148–51, 178
Foxglove Pug 137, 149, 151
Frosted Orange 140
Fylde Convention 2

Galashiels 164, 165, 180
Galium Carpet 151
Galtier, J. 200
Gametophyte 168, 169
Garden Tiger 90, 110
Garden Warbler 73
Gardner, F. W. 121, 161, 179
Gardner, John 117, 121
Gargunnock 159
Garrett, Dr. Frederick Charles 119, 161
Garside, A. 162
Gastrioceros 52
Gatekeeper 111, 114
Gavinton 13, 15, 83, 88, 101, 104, 124, 126, 128, 130, 131, 135, 140, 142, 143, 144, 146, 185, 187
Geikie, Archibald 20, 201, 202, 212, 213, 217, 226, 237, 238, 259
Genera Plantarum 108
Genomosperma 18, 172, 187

Genomosperma kidstonii 186, 187, 189, 190, 241
Genomosperma latens 183, 187, 189, 190, 212, 241
Gent, C. J. 161
Geology of Eastern Berwickshire 201, 212, 217, 226
Ghost Moth 261
Ghost Swift 109
Giant Bellflower 179
Gibb, Joseph 182
Gibside 110
Gibson, E. B. 60
Gibson, Samuel 61
Gibson, Thomas 61
Gilbert, O. L. 161
Gilly, Dr 209
Gilsland 165
Ginkgo 15
Ginkgo biloba 170
Glaucous Shears 131, 132, 148, 150
Glenroyd 9
Gnetopsis 197
Godwin, Dr. H. 242, 244
Gold Spangle 136
Gold Spot 135, 136, 139
Gold Swift 104
Golden Plover 86, 91, 145, 178
Golden Plusia 119
Golden Rod Brindle 84, 122, 125, 139, 140, 153
Goldfinch 82
Goldilocks 87
Goniatites 29, 37, 47, 52
Gordon Moss 24, 92–4, 100–04, 107, 124, 127, 129, 133, 134, 136, 138, 140–3, 145, 146, 148, 149, 152, 158, 180, 187, 242
Gordon, Prof. W. T. 195, 214, 215, 217, 230, 243, 258
Gorpley 29
Gothic 135, 152
Graham, Dr. 111
Grantshouse 13, 74, 104, 154, 156
Grayling 101, 141
Great Brocade 84, 116, 119, 125, 135, 137, 139, 141, 156
Great Water Beetle 136
Greater Swallow Prominent 134, 151

Green Arches 133, 135
Green Hairstreak 123
Green Sandpiper 79
Green Silver Lines 128, 150, 151
Green Woodpecker 71, 72, 85, 144
Green-veined White 74, 179
Green's Clough 29
Greenleighton Moor 117
Greenwood, A. D. 29–31
Grey Arches 113, 121
Grey Chi 154
Grey Dagger 116, 123, 133, 149
Grey Mountain Carpet 150
Grey Pine Carpet 113, 143
Grey Plover 77
Grey Scalloped Bar 127, 135, 148, 150, 151
Grey Spruce Carpet 113
Grey Wagtail 74, 75, 79, 91
Grime Bridge Pit 32
Grisedale Research Scholarship 53
Gullane 215, 258
Gunn, Rev. G. 175
Gunn, W. 213, 214, 217
Guthrie, W. G. 164
Gwynne-Vaughan, Prof. 20, 228, 232, 243
Gymnosperms 167, 169, 173, 207, 208, 233

Haggerston 118
Hairy Stonecrop 176
Hairy Violet 90
Halifax 4, 6
Halifax Hardbed Mine 48, 67
Halifax Infirmary 9
Hammersley, Mr. D. P. 180
Hancock Museum 16, 19, 24, 115–22, 160, 179, 180, 182, 193, 195
Hancock, Albany 110, 111, 160, 181
Hancock, John 110, 111, 160, 180, 181
Hanna's Bridge 15, 187, 189
Hardcastle Crags 123
Hardy, James 100, 102, 103, 112, 114, 118, 147, 175, 178
Hargreaves, H. 118

Harrison, Prof. J. W. H. 118–20, 161, 163, 181
Hart's Tongue Fern 87, 179
Hartley 111
Hartog, M. 219
Hartog, P. J. 219
Hartside 105
Haugh Head 179
Haughton 162
Haworth's Minor 155
Heath Rustic 140, 154
Hebclough 32
Hebden Bridge 47, 60, 123, 135
Hebrew Character 73, 85, 129, 130, 145, 148
Heddon on the Wall 162
Hedge Brown 111, 114, 155
Hedge Rustic 155
Hemingway, W. 43
Hemlock Water-Dropwort 141
Hen Hole 85
Henzell, Mrs. P. 181
Herald The 149, 150
Heterangium 59, 230
Hetton Hall 159
Hexham 119, 221
Hirmer, M. 232, 243
Hirsel 23, 92, 102, 104, 107, 146–9, 151, 152, 156, 181
Hoardweel 75, 103, 107
Hodgkin, T. E. 161
Holden, H. S. 232, 244
Holloway, Prof. 62
Holmes, J. C. 200
Holmfirth 7
Holoptychius nobilissimus 202
Holy Island 77, 180
Holywell 181
Hooker, J. D. 30, 69
Hooppell, R. E. 159
Horncliffe 159
Horne, John 228, 243
Hornet Clearwing 103
Hornshole 164
Horse Roads Bay 227
Hostimella 61
Hule Moss 78, 129
Hull 5, 7, 184
Hull, J. E. 118, 119
Humbie 180

Hume 107
Hume, Mr. John E. 175
Humming Bird Hawk Moth 88, 110, 142
Humshaugh 162
Hungry Snout 23, 24, 89, 126, 187
Hunt, Gladys 11
Hunt, Henry 11
Huntlywood 176
Hutton Bridge 234, 242
Hutton Bridge Mill 158
Hutton Mill 20, 22, 189, 191, 212, 232
Hutton, James 141
Hydrasperma 189, 191, 192, 198, 199, 201

Ingram Valley 117, 123
Innerleithen 164
Inskip 1, 4, 11, 24
Iris pseudacorus 179
Iron Prominent 140, 151

Jameson, Prof. R. 204
Jardine, Sir William 111
Jedwater 164
Jeddah 3
Jefferson, T. W. 163
Jeffrey, Alexander 227, 243
Jesmond 19, 113, 117
Johnson, John Robert 118–20
Johnston, Dr. George 75, 99, 100, 101, 111, 158, 159, 175, 176, 182, 210, 217
Joicey, James John 117
Joy, K. W. 69
Judd, Prof. J. W. 204, 218, 219
Juncus articulatus 178
Juncus bulbosus 178
Juncus squarrosus 178
Juniper Carpet 103, 156
Juniper Pug 149

Kalymma tuediana 94, 233, 241
Kelly, Andrew 89, 104–7, 123, 145, 158
Kelso 12, 165, 226, 242, 249, 257
Kendal 165
Kestrel 79, 85
Keswick 165

Kew 59
Kidston, Robert 20, 22, 43, 59, 94, 213, 214, 225–35, 238–41, 243, 257, 258
Kielder 122
Kielder Dam 162
Kiltorkan 170
Kingfisher 79
King Water 19, 165
Kirkwood, Rev R. 115
Knoll Hospital 13
Knot 77, 81
Knotgrass 131, 133, 140, 149
Knotted Pearlwort 176
Koopmans, Dr. 53
Krassilov 199, 200
Krausel & Weyland 232, 244
Kyles Hill 13, 24, 73, 80, 82, 84, 85, 92, 101, 124, 126–8, 131, 135, 136, 139, 140, 143–5, 148, 150, 151, 153, 155, 187
Kylins, Morpeth 181
Kyloe 118, 122

Lacey, W. S. 15, 69
Lady's Bedstraw 110, 115, 127
Ladykirk Burn 228, 240
Lagenicula 227
Lagenostoma 18, 51, 254, 258
Lagenostoma lomaxii 55, 254, 257, 258
Lagenostoma ovoides 15, 31, 33, 34, 44, 53, 54, 55, 58, 61, 63–5, 67, 69, 184, 189, 254
Laidlaw, W. B. R. 107
Lang, Prof. W. H. 20, 21, 29, 30, 32, 33, 45, 59, 61, 62, 65, 184, 185, 232, 256, 259
Langleeford 118, 120, 123, 180
Langmuir Moss 105
Langton Burn 15, 16, 20, 71, 74, 78, 87, 94, 107, 125, 186–8, 207, 227–32, 240, 241
Langton Glen 104, 188, 240, 242
Langtonlees Dean 177
Lanton 165
Large Argent and Sable 113
Large Blood Vein 122
Large Brocade 137
Large Ear 143

Index

Large Elephant Hawk Moth 70, 91, 104, 111, 112, 124, 149, 152, 180
Large Emerald 94, 136–9
Large Footman 115
Large Heath 91, 93, 111, 114, 117, 119, 136, 146
Large Nutmeg 113
Large Rivulet 132
Large Seraphim 122
Large Tortoiseshell 103, 115
Large Wainscot 126, 155
Large Yellow Underwing 110, 133, 139
Lartington 94, 220, 222, 223
Lathyrus odoratus 172
Lauder 80, 123, 181
Lauder and Lauderdale 158
Law, Robert 44
Lead Belle 121
Leaderfoot Bridge 159
Least Black Arches 132, 148
Least Yellow Underwing 122
Le Clerq, Prof. S. 69, 259
Leek High School for Boys 11, 184
Lees Cleugh 73, 80, 82, 83, 107, 124, 126, 127, 179
Legerwood 107
Leitholm 245
Lemmington 159
Lempke's Gold Spot 122
Lennel 159, 226, 257
Lennel Braes 94, 193, 206–8, 210, 213, 214, 216, 228, 229, 239, 240
Lepidocarpon 20, 37, 63, 65, 66
Lepidodendron 20, 38, 40, 194, 206, 211, 226, 228, 229, 239, 240
Lepidodendron brevifolium 59, 240
Lepidodendron nathorsti 214, 241
Lepidodendron rhodeanum 231
Lepidodendron spitsbergense 229, 230
Lepidodendron vasculare 33, 34, 37, 47, 49–51, 54, 57, 58, 62, 63, 65–8
Lepidodendron veltheimianum 229, 230, 240
Lepidophloios fuliginosus 33, 38, 48, 50, 51, 53, 57, 62, 63, 65, 66, 67
Lepidophyllum lanceolatum 212, 231

Lepidostrobus 31, 35, 46, 48, 51, 56, 66, 69, 229, 231, 239, 241, 244
Leslie, Rev. David S. 236
Lesser Butterfly Orchis 94
Lesser Cream Wave 113
Lesser Satin 149
Lesser Swallow Prominent 124, 131, 133, 137, 138, 147, 151
Levicaulis arranensis 240, 244
Lewes 10
Lewes County School for Boys 10, 184
Light Knotgrass 104, 151
Lightfield Farm 175
Lilburn Glebe 179
Lilburn Tower 159
Lime-speck Pug 151
Lindean Reservoir 164
Lindley & Hutton 193, 194, 201, 205, 216
Ling Pug 136
Linhope 117
Linnaea borealis 175
Linnaeus 108
Linnet 82, 89
Linum catharticum 178
Little, Adam R. 176
Little Auk 71
Littleborough 6, 36
Littlewood, Frank 165
Logan-Home, Lieut. Col. W. M. 101, 104, 105, 107, 122, 126, 181
Loligo forbesi 179
Lomax 36, 254, 255, 257
London Brindled Beauty 122
Long, Albert James 2, 3, 4
Long, Arthur Henry 3
Long, David Geoffrey 13, 159, 177
Long, Dora 3
Long, Edith Mary 3
Long, Frederick William 2
Long, Geoffrey 5, 6, 12
Long, George 2
Long, George William 2
Long, Gladys 11, 13, 129, 185, 253
Long, Helen Anne 3
Long, Jessie Naomi 1, 5, 6, 13
Long, Katie Louise 3
Long, Margaret Jean 13, 148, 152
Long, Marianne Bertha 5, 6, 11

Long, Marianne Elizabeth 3
Long, Samuel 2
Long-tailed Field Mouse 92
Longcroft Braes 106
Longformacus 107, 182
Longformacus Strip 176
Longframlington 181
Low Shilfurd 161
Lowther, R. C. 165
Lunar Hornet Clearwing 89, 112, 143, 145
Lunar Marbled Brown 124, 131, 147
Lunar Thorn 127, 131, 147, 151
Lunar Underwing 142, 154
Lunar Yellow Underwing 150
Lyginodendron 214, 215
Lyginopteris 44, 48, 50, 51, 53–5, 57, 58, 63, 65, 67, 68, 195, 215, 240, 255
Lyginorachis 240, 244
Lyginorachis papilio 195, 233, 240
Lyginorachis waltoni 240
Lyon, A. G. 20
Lyrasperma 189, 190
Lysimachia nemorum 179

Macconachie, Arthur 94, 227–31, 236, 237, 243
MacGregor, A. G. 202, 204, 217
Mackie, Dr. 22
Macnicol, Dr 130, 143, 145
Magnolia 192, 199
Magpie Moth 98, 109, 154
Maidens Blush 103
Makerstoun 180
Maling, William 115–17, 181
Mallow 103
Manchester Museum 62
Manchester Treble Bar 146
Manchester University 5, 15, 16, 20, 27, 28, 30, 65, 184
Marbled Brown 131, 149
Marbled Coronet 151
March Moth 71, 76, 85, 129, 144
Marchantites 228, 229
Marden 22, 24
Marsh Cinquefoil 94
Marsh Fritillary 100
Marsh Pennywort 94

Marsh Square Spot 120, 122, 127
Marsh Tit 71, 76, 77, 87, 139
Marshall Meadows Bay 230, 234
Maslen 256
Matheson, Adam 227, 228, 233
Mauldsheugh 164
Maxton 165
May Highflyer 73, 132
Mazocarpon shorense 66
McLean 254, 255, 258
Meadow Brown 93, 178, 179
Meadow Pipit 82
Meikle, Dr 164
Meldon Park 114
Mellerstain 107, 175
Melrose 226, 257
Mennell, H. T. 160
Mentha aquatica 178
Merlin 86
Miadesmia membranacea 66
Miller 102, 124, 133, 135
Miller, Hugh 31, 61, 170, 171, 183, 195, 202, 217, 235, 258
Millstone grits 27
Milne, David 208, 216, 227, 242, 243
Milne Home, Mr D. 208, 211
Minor Shoulder Knot 136, 138, 139, 153
Minsteracres 205, 221
Mitchell, Dr. G. F. 242, 244
Mitford, Bertram 111
Mitrospermum berwickense 190
Mitrospermum compressum 67
Molinia caerulea 178
Monynut 107
Moravian Brethren 7
Morebattle 165
Mother Shipton 132
Mottled Grey 78, 85, 129, 130, 144
Murchison, Sir Roderick 111
Muslin Ermine 179
Myeloxylon 67
Myosotis caespitosa 179
Myxomatosis 107, 137, 139

Nabb 6, 32–4, 37, 41, 44, 56, 57, 64–6
Narrow-winged Pug 133, 149

Natural History and Antiquities of Northumberland 108, 160
Needler Hall 8, 10
Neglected, The 124, 139
Nether Whitlaw Moss 164
Neuropteris 202
New Albany Shale 232, 244
Newham Bog 118
Nicholson, G. T. 119, 160, 163, 166
Nicol, William 95, 204, 209, 216, 218, 219
Ninebanks 118
Noctule Bat 10
Noeggerathia 230
Nomophila noctuella 140
Norham 159, 181, 209, 226, 257
Norham Bridge 20, 94, 210, 214, 227–30, 232, 233, 240, 241
Norman, Captain F. M. 177, 204, 220–3
North Tyne 122, 161, 162
Northern Dart 122
Northern Drab 85, 104, 130, 146
Northern Eggar 89, 133, 145, 178
Northern Naturalists' Union 119
Northern Rustic 90
Nosema 81
November Moth 120
Nowell, John 61
Nucellus 169
Nut Tree Tussock 130, 148
Nutmeg, The 154

Oak Beauty 85, 122, 129, 146
Oak Fern 96, 131
Oak Lutestring 117
Oakhill Clough 29
Old Cambus Quarry 153, 154, 155, 187
Old Hawick 164
Old Lady 141, 152
Old Meadows Colliery 31, 47, 48, 51, 54, 55, 57, 58, 61, 62
Oldham 63, 65, 66, 68
Olive, The 154
Oliver, F. W. 16, 36, 204, 226, 254–8
Ootheca globosa 234, 239
Ophioglossum vulgatum 32, 34, 37, 38

Orange Sallow 126, 142
Orange Swift 137
Orange Tip 100, 101, 103, 104, 108, 111, 158–66, 180
Ornsby, Rev. George 162
Orthocerata 210, 211
Ovens, Thomas Middlemiss 234–9, 243
Ovule 169, 172–4, 191, 197–200
Oxendean 24, 129, 130, 141, 142, 146
Oxroad Bay 19, 174, 180, 191

Painted Lady 96, 97, 111, 158
Pale Brindled Beauty 10, 71, 75, 83, 128, 144
Pale Oak Eggar 136–9, 153
Pale Prominent 124, 132, 141, 149, 151, 152
Pale Tussock 105, 145
Pale-shouldered Brocade 132, 148
Paravespula 181
Parrack, J. D. 112, 122, 161, 179, 181
Parsley 90
Pate, Mrs. E. O. 175, 176
Paterson, Mr. D. 102
Paxton 104, 158
Peach Blossom 135, 136, 149, 151
Peach, B. N. 213, 214, 227
Peach, Charles 61, 227
Peacock 109, 111
Pearl Bordered Fritillary 96
Pearly Underwing 88, 116, 142, 143
Pease Bay 89, 141, 203, 227, 259
Pebble Hook-tip 132, 140, 150
Pebble Prominent 133, 151
Pedicularis sylvatica 178
Peebles 164
Pelham-Clinton, E. C. 102, 130, 143, 145, 157, 165, 166
Pencaitland 180
Peppered Moth 131, 149, 151
Peregrine 86
Pettit J. M. 170, 171, 183
Pettycur 43, 59, 60, 184, 186, 233, 242
Phyllitis scolopendrium 179
Physostoma 257

Pied Flycatcher 73, 86, 130
Pied Wagtail 74
Pimpinella saxifraga 179
Pine Beauty 73, 85, 128, 146, 148
Pine Carpet 136, 137, 139, 153
Pinites brandlingi 194
Pinites medullaris 193–5
Pinites withami 194, 195
Pink Barred Sallow 142
Pink-footed Goose 77, 78, 85, 128
Pitus 19, 190, 195, 213–5
Pitus antiqua 94, 193, 208, 214, 233
Pitus primaeva 193, 195, 208, 233
Pitus Withami 193, 194, 207
Pitys dayi 215
Plain Wave 113
Plashetts Pond 161
Plasmodium 25
Plessey Woods 160
Pleurotomaria 210, 211
Pochard 78
Polson, H. 159
Polwarth 245
Ponteland 15, 19, 113
Poplar Grey 102
Poplar Hawk Moth 91, 111, 149
Poplar Kitten 102, 121
Populus tremula 179
Porteous, A. M. 104
Portland Moth 103
Portsmouth 108
Posidonomya 52
Potamogeton polygonifolius 178
Powburn 160
Powdered Quaker 102, 146, 147
Prankerd, Miss T. L. 56, 69, 254
Presidential Address to B. N. C. 1972 167–75
Preston 13, 22, 70, 103, 123
Preston Haugh 170, 202
Preston Schoolhouse 13, 185
Prestwick Carr 110, 111, 114, 118, 120, 122
Pretty Pinion 122
Prickly Shield Fern 87
Prothallus 168, 169
Protoclepsydropsis 232
Prunella vulgaris 178
Psalixochlaena berwickense 200
Psalixochlaena cylindrica 67

Pteridosperms 16, 19, 169, 170, 172–4, 192, 195, 197, 199, 200, 227, 233, 238–40, 255
Pterichthys major 202
Pterinopecten 47, 52
Purple Bar 127, 133
Purple Clay 133, 134
Purple Hairstreak 104
Puss Moth 110
Pybus, Mrs 162

Quail 135
Quaking Grass 179

Rachiopteris aspera 240
Rachiopteris multifascicula 233
Raistrick, A. 42, 69
Ramsden Clough 35, 52
Ranunculus acris 178
Ranunculus flammula 178
Raven 86
Read, C. B. 232, 233, 244
Red Admiral 108, 111
Red Carpet 136, 138, 139
Red Chestnut 73, 85, 130, 131, 147
Red Green Carpet 143
Red Line Quaker 142, 156
Red Sword Grass 106, 116, 119
Red Twin-spot Carpet 133, 134, 148
Red Underwing 10, 102
Reddish Light Arches 116
Redesdale 119, 120, 161
Redheugh Wood 161
Redpath Moss 107
Redstart 96
Reed Bunting 90, 145, 148
Reid, Rev John 236
Renton, Robert 104, 107, 123, 158, 178
Renton, William 100
Reseda luteola 173
Rest Harrow 90
Reston 104, 201
Retreat 23, 83, 84, 128, 129, 131, 135, 137, 141, 143, 149, 187
Rheims 12
Rhetinangium arberi 230, 233, 243
Rhizodus hibberti 211
Rhodea moravica 230, 235

Rhynie 20, 22, 232
Riddell, Rev. H. 182
Ridge Bank 5, 34
Riding Mill 121, 160–2, 179, 221
Ringed Plover 77, 82
Ringlet 93, 136
Rivulet 150
Robinson, J. R. 161
Robson, J. E. 116, 117, 160, 179, 181
Robson, J. Percy 120
Rock Pipit 95
Rockrose 90
Roman Wall 19
Roomfield 1, 4, 7
Rose Bay Willow-herb 70, 104, 180
Rosie, Alexander 121
Rosie, Annie 121
Rosie, David 121
Rosie, Donald 122
Rosy Minor 137
Rothbury 181, 193
Rothwell, Mr. 41, 43
Rough Bank Colliery 45
Rowe, Thomas 45
Roxburgh 164
Roxburgh Castle 180
Royal Holloway College 60
Ruby Tiger 89, 144, 145
Ruddy Highflier 104

Sabden Shales 29
Sagenaria Veltheimiana 212
Sagina nodosa 176
Salad Burnet 90
Salisbury 255
Salix 193, 199
Salix aurita 179
Salix herbacea 242
Salix reticulata 242
Sallow Kitten 132, 149, 151
Salpingostoma 190
Salter, J. W. 212
Samaropsis bicaudata 229, 241
Samaropsis scotica 241
Samieston 164
Sand Dart 122
Sanderling 82
Sanderson, Mr. 204, 209

Sanderson, W. 243
Sandpiper 96, 179
Satellite, The 73, 145
Satin Beauty 121, 122
Satyr Pug 133
Saxifraga hirculus 176–8
Saxon, The 133, 135, 149, 151
Scalloped Hook-tip 124, 150
Scallop Shell 121, 143
Scarce Bordered Straw 103
Scarce Prominent 120, 121, 147
Scarce Silver Y 136, 137
Scarce Tissue 107, 156
Scaumenac Bay 170
Scaup 82
Schistostega 64
Scholes 7
Scirpus caespitosus 178
Scorched Wing 122
Scotch Argus 100, 103, 104, 123
Scotch Brown Argus 113, 123
Scot's Gap 245
Scott, D. H. 16, 30, 36, 42, 187, 193, 194, 204–6, 214, 215, 217, 218, 221–5, 228, 231, 232, 240, 243, 254–7
Seaton Sluice 179, 181, 182
Sedum villosum 176, 178
Selaginella 168
Selby, P. J. 101, 102, 111–13, 124, 159, 181
Selkirk 164
September Thorn 103, 141, 142
Setaceous Hebrew Character 133
Seward, A. C. 204, 217, 219, 231, 232, 243
Shackleton 8
Shaftoe Crags 162
Shag 82
Sharneyford 37, 56
Sharp-angled Carpet 122
Shaw, William 100, 102, 103, 105, 123, 158
Shears, The 150, 151
Sheepscombe 2
Shelduck 82
Sheppard, D. A. 122, 157, 162, 179
Shiningpool Moss 178
Shore 36, 38, 46, 55–8, 66, 257
Short-eared Owl 89, 138, 145, 179

Shoulder Stripe 130, 146
Shoulder-striped Wainscot 135
Shoveller 78
Siccar Point 88, 90, 141, 181
Sidwood 161, 179
Sieglingia decumbens 178
Sigillaria 29, 34, 48, 50, 53, 57, 58, 66, 67, 206, 210, 212
Silky Wave 113
Silver Y 96, 97, 151, 158
Silver-striped Hawk Moth 97, 102
Silver-studded Blue 101
Silver-washed Fritillary 104
Silvertop 205, 220, 221, 223
Simonburn 160, 181
Simpson, F. 69
Siphogamy 173
Siskin 76
Slaggyford 161
Slaley 162
Slender Pug 138
Small Angle Shades 131, 133
Small Argent and Sable 91, 133
Small Autumnal Carpet 120, 141, 153
Small Blue 111, 112, 148
Small Chocolate Tip 112, 132
Small Clouded Brindle 103, 149
Small Copper 52, 54, 108
Small Eggar 112
Small Elephant Hawk Moth 91, 151
Small Engrailed 118
Small Fanfoot 126, 150–2
Small Fan-footed Wave 138
Small Heath 52, 91, 93, 148, 178
Small Mottled Willow 117
Small Pearl-bordered Fritillary 93
Small Phoenix 132, 140, 153
Small Quaker 73, 85, 128, 129, 146
Small Square Spot 122, 133
Small Tortoiseshell 96, 109, 153, 178
Small Yellow Underwing 104
Smith, A. J. 159, 164
Smoky Wave 119
Sneap 161
Snipe 77
Snow Bunting 82, 89
Solms-Laubach, Count 231, 243
Sorby 204, 218

Sourhall 56
Spadeadam 19
Sparrowhawk 79
Species Plantarum 108
Speckled Wood 101, 104, 107, 112, 123
Speckled Yellow 122
Spencer, James 44
Spencerites insignis 66
Sphenophyllum 34, 46, 51, 54, 58, 68
Sphenopteridium pachyrrachis 235
Sphenopteris affinis 195, 239
Sphenopteris bifida 198, 201
Sphenopteris dissecta 231
Sphenopteris elegans 229, 235
Sphenopteris pachyrrachis 235
Sphenopteris patentissima 228
Spirorbis 210, 239
Spittal 181
Spitzbergen 64
Sporophyte 168, 169
Spottiswoode 104, 107, 135, 136
Sprawler, The 121
Spring Usher 76, 83, 128, 144
Square Spot Rustic 137, 139
Square-spotted Clay 103, 116
Squid 179
St. Abbs Head 152, 179
St. Boswells 12, 165
St. Mary's Loch 164
Stag's Horn Moss 87
Stamfordham 160
Stamnostoma 18, 22, 189, 190, 195
Stansfield, Abraham 61
Stark, Rev. I. 204, 221
Stauropteris 45, 51, 54–6, 58, 65, 66, 68, 171
Stauropteris burntislandica 43, 59, 183, 213, 242, 244
Staward Peel 161
Stebbins 197, 201
Stenomyelon 20, 187, 190, 227, 230, 233, 244
Stenomyelon muratum 232
Stenomyelon tripartitum 232
Stenomyelon tuedianum 227, 232, 233, 241, 243
Stephanospermum 255
Stevenson, Mr. 211, 212, 216
Stewart, John 202

Index

Stewart, W. N. 197, 201
Stichill 159
Sticks, Harry 120
Stigmaria bacupensis 51, 57, 62, 63, 66
Stigmaria ficoides 51, 57, 58, 63, 66, 205, 210, 228–30
Stigmaria lohesti 66, 69
Stirling 226
Stobtree Whin 118
Stockmans, Dr. 52, 53, 259
Stockport 2
Stocksfield 161
Stonechat 90, 95
Stonecrop 90
Stoodley Pike 260
Stopes, M. C. 69
Stout Dart 116
Stranraer 165
Straw Point 122
Streamer, The 132
Striped Twin-spot Carpet 113, 137
Stuart, Dr. C. 158, 175, 177, 182, 211, 217
Stubblefield, C. J. 69
Sturrock, Mr. 175
Suez 3, 88
Summers, P. 159, 160, 162, 180
Sundew 91, 178
Surange K. R. 172, 183
Suspected 126, 138, 153
Swallow Prominent 124, 134
Swallowtail Moth 152
Swan, Mr. Clayton 116
Sweet Gale 113
Swinton Mill 158
Systema Naturae 108

Tantallon Castle 19, 258
Tantallosperma 190
Tate, George 210, 211, 217
Tawny Barred Angle 151
Tawny Owl 80
Teal 78
Telangium affine 231, 235
Testimony of the Rocks 171, 183, 195, 202, 217, 235
Tewkesbury 3
Thalictrum alpinum 242
Thirlestane 164

Thomas, H. H. 198, 201
Thomson, A. 158
Thomson, Dr. R. D. 207, 208, 216
Thornton 159
Threeburnford 104
Threepwood 105, 143
Threeshire Heads 12
Thursophyton 61
Thyme Pug 119
Tinlin, Dr. C. 164
Tissue, The 107, 119, 132
Todmorden 1, 4, 6, 12, 18, 25, 27, 44, 99, 141, 157, 260
Todmorden Edge 157
Torkington Hall 2
Torreya californica 254
Toxteth Tabernacle 5
Traquair, Dr 236
Treble Bar 119, 136–8
Treble Lines 119
Tree Pipit 73
Tree Sparrow 82, 83
Trevelyan, Sir John 111
Trigonocarpon 30, 46, 69
Triletes crassiaculeatus 241
Trimerophytopsida 200
Triple-spotted Clay 117, 123
Tristichia ovensi 195
Trochodendrocarpus 199
Tufted Duck 78
Tufted Forget-me-not 179
Turnbull, John 106
Turnip Moth 149, 156
Turnstone 82, 95
Tutin, T. G. 64
Tweed Mill 193, 206, 208–10
Twin-spot Carpet 178
Twizel 101, 112, 113, 124, 159, 181
Tynan, A. M. 162
Tynemouth 110, 111
Tynron 165

Ulodendron 40
Uncertain, The 112
Ushaw College 193

Valerian Pug 118
Vapourer, The 103
Velvet Scoter 82
Veronica officinalis 178

Veronica scutellata 178
Vestal, The 158

Wailes, George 113, 114, 159, 160, 162, 181
Waldie, J. & G. 175
Wall, The 123, 181
Wall Brown 105, 107, 108, 112
Wallace, I. & B. 122
Wallington 160, 162
Wallis Club 114, 119
Wallis, John 30, 69, 108, 109, 160, 162, 181
Walton Cardiff 2
Walton, Prof. John 16, 30, 69, 192, 193, 204, 218, 225, 231, 244, 255
Warden 162
Wardle 35, 36, 55
Wark 161
Wasserman, John G. 115, 117
Watch Valley 107
Water 6, 56
Water Boatman 136
Water Carpet 130
Water Rail 74, 96, 138
Waterston, Dr C. D. 19, 213
Waterston, Dr. G. 180
Watson D. M. S. 69
Watson, Mr. C. 177
Watson, W. G. 113, 161
Watsonian vice-counties 100
Weak Law 215
Weiss, Prof. 45, 69
Wellcleugh Burn 176, 178, 179
Wells Sawmill 165
Welsh Wave 127, 135, 137, 139, 151
West Blanerne 20, 22, 157, 174
West, Mr. P. W. 180
Wheatear 96
Wheeler, R. F. 159, 160
Whitburn 242
Whiteadder 16, 20, 22, 23, 75, 80, 94, 101, 158, 170, 171, 173, 174, 180, 185, 191, 197, 203
White Colon 116
White, E. I. 235, 238, 243
White Ermine 109
White-line Dart 137, 155

Whiterigg 242
White Satin 102, 112
White Underwing 122
White Wagtail 90
Whitley Bay 179, 181
Whitley Castle 108
Whitrig Bog 242
Whittle Dene 162
Wickenberry Clough 25
Wideopen 194
Wigeon 76, 77
Wight, Marianne Byam 2
Williams, C. B. 97
Williamson, Prof. W. C. 30, 204, 219, 226, 254
Williamstone 161
Willie's Hole 228, 229, 241
Willis, A. J. 69
Willow Burn 214
Willow Warbler 179
Winch, N. J. 205, 208, 216
Winlaton Mill 110, 164
Winter Moth 76
Witham, Henry T. M. 22, 94, 95, 193–5, 203–11, 213, 215, 216, 218–25
Witham, Philip 204, 220–4
Witham, Rev. T. 220–3
Wood, M. 61
Woodburn 257
Woodcock 75, 176
Wood Vetch 90, 156
Wood Warbler 73, 130
Wormwood Pug 152
Wright, Dr. W. B. 34, 40, 42, 69
Wylam 181

Xanthoria parietina 90

Yellow Horned 85, 90, 128, 130, 144, 145
Yellow Iris 179
Yellow Marsh Saxifrage 176–8
Yellow Pimpernel 179
Yellow Tail 115

Zalessky 214
Zeilleria moravica 235
Zygopteris kidstoni 232, 233